EROSION AND SEDIMENT YIELD:

SOME METHODS OF MEASUREMENT AND MODELLING

Edited by

R. F. HADLEY AND D. E. WALLING

Contributors

H. A. Elwell

W. W. Emmett

R. F. Hadley

H. G. Heinemann

C. A. Onstad

D. L. Rausch

D. E. Walling

P. R. B. Ward

published by:

Geo Books,
Regency House,
34 Duke Steet,
Norwich NR3 3AP,
England.

ISBN hardback 0 86094 158 2

ISBN paperback 0 86094 141 8

Printed in Great Britain at the
University Press, Cambridge

CONTENTS

PREFACE

Processes of erosion, transport, and deposition of either weathered rock or soil are part of the normal cycle that shapes the fluvially eroded landscapes of the Earth. The rates at which the processes act are dependent on such variables as rock or soil type, topographic relief, plant cover, climate, and land use. Human activities, which influence the land-use variable, have a marked effect on all processes in the cycle. In order to assess environmental problems related to land use, empirical methods have been developed to measure rates of erosion, transport, and deposition.

The primary purpose of this book is to present some methods of measuring erosion and sediment yield of solid matter and solutes in a variety of environments and land uses. Some methods are also presented for utilizing erosion and sediment yield data in the development of regression equations and models designed to improve predictive and estimation capabilities.

The relation of erosion on an upland site to sediment yield at a measurement point downstream in the drainage network has been the subject of considerable research in the past four decades. Progress in recent years has resulted in a better understanding of erosion processes and methodologies for routing products of erosion through a drainage system. An ultimate goal, however, is a predictive sediment yield model that is based on physical processes. Empirical relations, such as sediment-delivery ratios do not satisfactorily explain the processes. If it is assumed that a part of the eroded material from upland hillslopes accumulates as colluvium and floodplain deposits, then the concept of equilibrium in sediment transport cannot be valid. The storage of sediment en route in the system must fluctuate with changes in land use, conveyance efficiency of channels, and climatic conditions. As Wolman (1977) so aptly stated, the sediment-delivery ratio provides a cover for physical-storage processes and errors in estimates of erosion. Before we can develop sediment yield models, and a better understanding of the linkage between erosion and sediment transport, we must improve our observational data base and process understanding.

The individual chapters assembled in this book include: (1) simple methods of measurement of upland erosion and sediment yield; (2) qualitative methods of estimating sediment yield; (3) methods of measurement of suspended load and bedload of streams; (4) measurement of dissolved loads in streams; (5) methods of measurement of reservoir

v

sedimentation and trap efficiency; (6) empirical
estimates of upland soil loss; and (7) modelling of
sediment yield. It is not possible in a single volume,
such as this one, to consider all the empirical methods
and modelling procedures that have been developed to
solve erosion and sediment yield problems. The methods
presented, however, are representative of current
research in the field of erosion and sediment transport.

The instrument technology for measuring and
recording streamflow, suspended sediment, and chemical
constituents has been developed to a high degree of
sophistication and installed at gauging stations on large
rivers in many countries. Instrumentation for measuring
hillslope erosion processes and hydrologic parameters has
been improved in recent years, but it is generally used
only on selected experimental watersheds. This further
complicates the problem of routing sediment from upland
hillslopes to gauges on large drainage basins because of
the scarcity of data from small basins.

This book is a contribution of the International
Commission on Continental Erosion (ICCE) of the
International Association of Hydrological Sciences.
The Commission maintains a close liaison with governmental
and nongovernmental organizations, universities, and
individuals throughout the world on current research in
erosion and sedimentation processes. We hope that the
methods described in this book will be useful, especially
to those scientists and engineers who are initiating
research and data-collection programmes concerned with
erosion and sedimentation problems.

R. F. Hadley

D. E. Walling

1.

Measuring and predicting soil erosion

R. F. Hadley

Introduction

Soil erosion and the problems related to it, such as loss of soil productivity and damage caused by deposition of transported sediment, have been a matter of concern for hydrologists and agricultural engineers for several decades. Geomorphologists have studied erosion and sedimentation because of their interest in the processes that shape hillslopes and drainage networks that comprise fluvial landscapes. In order to understand the processes of erosion, field methods, both simple and sophisticated, have been developed to measure the quantity of soil that is eroded in a specific period of time. These measurements, together with data on precipitation characteristics, soil and vegetation characteristics, topographic characteristics, and land use can be used to develop empirical relations for prediction of soil erosion and estimation of sediment yield.

Simple measurement methods

Field experiments designed to measure soil erosion are generally conducted on small plots or drainage basins less than 5.0 km^2 in size. This is done partly to reduce the cost of instrumentation, and partly to facilitate the task of data collection. When the area of the field experiment is small, it is easier to maintain close control of the accuracy of measurements and to observe hydrologic response to individual runoff events. Monetary and manpower constraints, however, generally make it impractical to install instrumentation on a large number of small upland drainage basins. Therefore, simple reconnaissance methods have been developed to measure erosion and sediment yield on basins of 2.5 to 5.0 km^2. Simple techniques for monitoring sheet and gully erosion on hillslopes, channel aggradation and degradation, mass movement of soil on steep slopes, and channel geometry changes have been developed in the past 30 years. (Emmett, 1965; Emmett and Hadley, 1968; Dunne, 1977).

Hydrologist, US Geological Survey, Water Resources Division, Box 25046, Denver Federal Center, Mail Stop 420, Lakewood, Colorado, USA.

Several simple measurements of morphological changes on hillslopes and in stream channels that do not require elaborate field equipment are described in the literature (Miller and Leopold, 1963; Hadley, 1965; Emmett, 1965; Campbell, 1974). Although the basic data on geomorphic processes that are collected by using these techniques are useful, they generally are useful only in evaluating conditions and trends during long periods of time (Hadley, 1965; Emmett and Hadley, 1968).

Measurements of surficial erosion and mass movement on hillslopes may be made using any of several simple techniques. Establishment of slope transects with a series of steel pins or nails have been frequently employed. The amount of erosion at each pin can be measured in two ways. The pin may be left exposed a known distance (generally 6.0 cm) above the ground surface and the length of pin freshly exposed can be measured at each visit. Another way of measuring erosion is by using a long nail and a washer that fits against the head of the nail. These are driven into the slope flush with the ground surface and the amount of erosion is measured by the distance that the washer slides down the shaft of the nail and it is undermined (Miller and Leopold, 1963). There has been some questions raised by researchers about the accuracy of this type of observation (Schumm, 1967) because of the disturbance of the surface during installation and measurement, raindrop impact, and the increased roughness caused by the pin that may cause scour. These objections may be valid in part, but the technique has been tested at many locations and it is believed that they provide reasonable estimates of erosion. Measurement of erosion-pin exposure also have been compared with sediment-yield data in a small drainage basin to test the reliability of the method.

Badger Wash basin in western Colorado, near Grand Junction, is the site of the long-term erosion and sediment-yield studies by the U.S. Geological Survey. Erosion pins were installed on hillslopes in a small basin (4.8 ha) in the Badger Wash basin. There is a reservoir at the mouth of the basin where runoff and sediment yield are measured. Hadley and Lusby (1967) measured the net erosion on six erosion-pin transects and the sediment yield delivered to the reservoir as a result of a thunderstorm on August 12, 1964. Because of the small size of the basin, steep slopes, and intricate dissection, there is little opportunity for intermediate deposition between the point of entrainment on the hillslopes and deposition in the reservoir.

The results of the research by Hadley and Lusby (1967) showed that approximately 134.16 m^3 of sediment was eroded from the basin during the storm and approximately 109.77 m^3 was delivered to the reservoir. This is a sediment-delivery ratio of 0.82, which is much higher than could be expected from a basin with diverse topography where deposition might occur at slope bases and in floodplains. The data acquired in this single runoff event are insufficient to make general statements about the accuracy of erosion-pin measurements, but they do indicate that such measurements are useful in estimating erosion rates.

Mass movement of the soil mantle can be measured by drilling rows of holes, generally less than 12.5 mm in diameter, in the soil to a depth of 0.9 to 1.2 m and normal to the slope and along a contour (Hadley, 1967). These holes may be filled with glass beads 1 to 2 mm in diameter. Excavation of the site at a later time will show any downslope movement of the soil mantle in cross section (Fig 1.1).

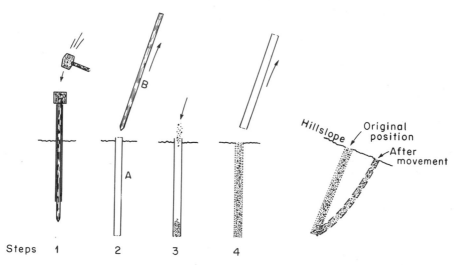

Steps   1   2   3   4

1 : With rod inserted in tube and upper end fitted with driving head, tube and rod are driven into the ground
2 : Withdraw rod
3 : Fill tube with colored grains
4 : Withdraw tube

Showing the type of downslope movement that might be expected

Fig 1.1.   Sketch of method for using glass beads to monitor mass movement

Other techniques for measuring erosion and mass movement are described by Miller and Leopold (1963) and Dunne (1977). Techniques for monitoring morphological changes in stream channels similar to those described for hillslopes can be easily installed at little expense. The simplest technique is the surveyed cross section between permanently monumented end-points (Emmett, 1965). The channel cross section can be surveyed annually or at any desired interval to monitor erosion, aggradation, and changes in channel position in the floodplain. These data combined with erosion data from uplands provide an approximation of erosion rates and distribution of source areas within a basin (Leopold, Emmett, and Myrick, 1966).

3

Channels of most rivers will scour their beds during periods of high flow and backfill on the flood recession (Emmett, 1965). The data from a monumented cross section surveyed annually generally will not furnish information on scour and fill during a single flood event. A simple technique has been developed (Leopold, Emmett, and Myrick, 1966) for collecting scour and fill data. A channel cross section is surveyed and monumented on both banks and a series of vertical holes are dug into the channel bed to a depth about twice the depth of expected scour. Leopold, Emmett, and Myrick (1966) used a depth of 1.2 m in ephemeral stream channels in New Mexico and had few losses of installations. A chain is placed vertically in each hole and firmly anchored with a rock at the bottom; the hole is then backfilled leaving a few free links at the surface. When a flow event occurs that causes channel scour, a length of the vertical chain will be laid over in a horizontal position, and on the recession sediment will be deposited over the horizontal chain. Resurvey of the cross section and excavation of the chain will produce data on net scour and fill similar to those shown in Fig 1.2 (Leopold, Emmett, and Myrick, 1966). This method does not tell the investigator at what time during the flow event scour occurs but it does provide information on net scour and fill.

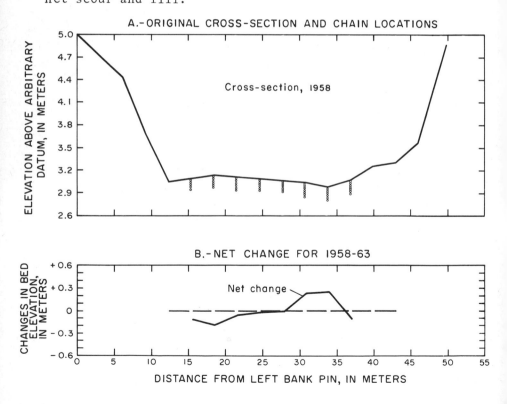

Fig 1.2. Channel scour chain installation and net change in bed elevation, 1958-63 (from Leopold, Emmett, and Myrick, 1966).

Use of small reservoirs

In small basins where reservoirs have been constructed,
data can be collected at the reservoir on runoff and
sediment yield by using the reservoir as a gauging station.
These hydrologic data can be used in conjunction with
simple techniques described for measuring hillslope and
channel erosion and information on drainage-basin
characteristics from maps or aerial photographs. The
relations between hydrologic and morphometric parameters
that can be developed from this type of data make it
possible to generally characterize basins on a regional
scale in areas of similar soils and climate. Differences
in runoff, sediment yield, and morphometry attributable to
rock and soil types, relief, and vegetation cover also are
apparent from analysis of these data (Hadley and Schumm,
1961).
These techniques and the resulting data that are
collected by using them are not substitutes for some of
the sophisticated instrumentation that has been developed
in recent years. The results obtained generally represent
long-term averages of erosion and sediment yield and cannot
be used in physical process models linked to individual
stormflow events. However, in areas where the data base
is meagre, these methods are useful.

Prediction of soil loss

Techniques for predicting soil loss have evolved over
the past four decades as research in soil-erosion processes
has increased. The simple equations became more complex as
more data became available, and the Universal Soil Loss
Equation (USLE) is now the most widely used equation for
soil loss prediction (Wischmeier and Smith, 1965).
The USLE includes all of the major causal and
conditional factors that affect the amount of erosion from
a specific site. It is an empirical equation that was
developed to estimate erosion losses from cropland. The
equation was basically developed from data collected from
small runoff plots located east of the Rocky Mountains in
the United States, but has been greatly extended in use.
Research needs for further extension and refinement in
different areas have been suggested by Singer, Huntington,
and Sketchley (1977) for California rangeland; El-Swaify
and Dangler (1977) for tropical soils; Aina, Lal, and
Taylor (1977) for western Nigeria rainforest; Roose (1977)
for West Africa; McCool, Molnan, Papendick, and Brooks
(1977) for the dryland grain region of the Pacific Northwest;
Osborne, Simanton, and Renard (1977) for the semiarid
southwestern U.S.; Evans and Kalkanis (1977) for general
use in California; and Brooks (1977) for Hawaii. Elwell
(Chapter 2, this book) has modified the USLE for conditions
in Zimbabwe. In addition, it has been extended for use in
urban areas, forest harvesting, roads and highways, mining
areas, and recreation areas where extensive areas of
subsoils may be exposed.

The USLE, as documented by Wischmeier and Smith (1965), to define soil loss is a series of factors:

$$A = R\ K\ L\ S\ C\ P$$

where A is soil loss in tons per acre or metric tons per hectare. R is an index of the erosive forces of rainfall and runoff (Wischmeier and Smith, 1958). It is the summation of the products of the kinetic energy and the maximum 30-minute storm rainfall intensity occurring on an annual basis. Where point rainfall data are inadequate to compute R by this method, it can be estimated on the basis of the expected 2-year 6-hour rainfall for the area. K is a factor representing the inherent erodibility of a soil. It was derived as the average annual soil loss per unit of factor R from a slope 72.6 ft (22.1 m) long and 9 percent gradient in an area of clean-tilled soil. L is the slope length factor and is the dimensionless ratio of soil loss from the field slope length to the soil loss from a slope 72.6 ft (22.1 m) long under identical conditions. S is the slope steepness factor and is the dimensionless ratio of soil loss from the field slope gradient to the soil loss from a 9 percent slope under identical conditions. C is the cover and management factor and is the ratio of soil loss from an area with specified cover and management to an identical area of untilled, continuous fallow soil. P is the factor representing benefits from the use of conservation practices expected to reduce erosion. It is the ratio of soil loss with a conservation practice to the soil loss with straight row farming up and down the slope. Detailed explanation for determining each of the factors are given by Wischmeier and Smith (1965, 1978).

The limitations of the USLE have been thoroughly analyzed by Wischmeier (1976). Some of these are included in the following:

1. Field sediment yield is usually much less than the sum of the soil losses from the several slope segments in the field because of deposition in depressions within the field, at the toes of slopes, along field boundaries, and in terrace channels, if any.

2. The equation computes long-term average soil losses; and actual amounts for individual periods of time can therefore be expected to deviate from this average by a wide margin.

3. The potential for considerable interaction exists among the different factors; for example, the cropping system with the rainstorm distribution. Indications of some interactions between soil, topography, and surface conditions require further research.

4.   The equation cannot be correctly applied to a complex watershed by using overall average factors for K, C, and LS -- extensive subdivision of the watershed into small parts would be necessary. Furthermore, in extension of item 1, the sediment yield of this complex watershed is much less than that obtained from the sum of the subdivided parts -- a sediment-delivery ratio is needed.

5.   If the equation is used to estimate the soil loss for a specific design rainstorm, the results would be an average for the expected conditions. However, soil losses for specific storms my be greatly affected by wide fluctuations in secondary parameters that temporarily affect the equation factor values. For example, some rains are accompanied by high winds and others occur in a calm atmosphere which greatly affects the impact energy of the raindrops; or, some rains begin with a very high intensity and quickly seal the surface so that the latter part of rain encounters a lower than normal infiltration rate.

6.   Computed results from hurricane-associated storms on relatively flat slopes are usually much too high. This may be attributed to shielding of the soil surface by ponding rainfall.

Rainfall simulation

In order to obtain adequate input to soil-loss prediction equations, such as the USLE, the data-collection process may take several years, especially in arid and semiarid regions where runoff events are frequent. Rainfall simulators offer the advantages of collecting data rapidly and being able to control the duration and intensity of rainfall (Meyer, 1965). For many years portable infiltrometers that utilize artificial rainfall have been used by investigators to collect data on infiltration, runoff, and sediment transport. Although these infiltrometers are portable and apply water uniformly to a small plot, there are some disadvantages. Boundary effects and the effects of isolated surface features generally are greatly magnified in the resulting data (Lusby, 1977).   In order to obtain results that more closely reflect natural conditions, a rainfall simulator was developed by the US Geological Survey that can be used on small drainage areas of about 370 m^2 (Lusby, 1977).
The US Geological Survey (USGS) rainfall simulator is patterned after a rainfall-runoff facility developed and installed at Colorado State University (CSU) (Holland, 1969). The major differences between the two simulators are: (1) the USGS simulator was designed to be taken into the field and operated with a water truck while the CSU facility is stationary; and (2) the USGS simulator operates at a single rainfall intensity of about 51 mm/hr regulated by pressure valves and the CSU facility has the capability of varying rainfall intensity from 13 to 108 mm/hr by controlling input to various sets of sprinklers regulated by valves. Both rainfall simulators basically consist of a grid of sprinker heads mounted on riser pipes 3.1 m high (see Fig 1.3).

Fig 1.3. Photograph of rainfall simulator in field operation, northern Nevada, USA.

Rainfall produced by this simulator probably has fewer large-size drops than natural rainfall especially at higher intensities, but this deficiency is partially offset by the uniform coverage on the basin (Lusby, 1977).

The USGS simulator requires 10,000 gallons ($37.8m^3$) of water, supplied from a truck, to produce an artificial storm of 38 mm in 45 minutes on an area of 370 m^2. Runoff from the basin is measured with a Parshall flume with a 25.4-mm throat, as is shown in Fig 1.4.

Fig 1.4. Photograph of Parshall flume used to measure runoff from rainfall simulator plot. Jars are used for sediment samples.

Measurements of stage are made at 1-minute intervals.
Samples are taken at 3-minute intervals at the outfall of
the flume for determination of sediment concentration and
chemical quality analyses.
The simulator has been used at several sites in the
western United States to characterize various soil-vegetation
communities and land-use practices with good results.
Extrapolation of results to larger areas, however, will
require measurements in the different hydrologic units that
make up the drainage basin (Lusby, 1977). Many types of
rainfall simulators have been developed and used to collect
hydrologic data in the laboratory and in the field in recent
years. An excellent discussion of the specifications of
some designs may be found in de Ploey and Gabriels (1980).

Estimation of sediment yield

The relation of the quantity of erosion on an upland
plot or small drainage basin to the sediment yield at some
measurement point downstream is not well-defined. Sediment
yield has been estimated using a variety of empirical methods.
Until the linkage of erosion on upland slopes, the storage
of sediment en route and the sediment transport processes
in streams are more completely understood, a physically-based
sediment yield model cannot be developed. There are, however,
methods for qualitatively estimating sediment yield based on
drainage basin characteristics and reservoir sedimentation
data.
A simple qualitative method for evaluating sediment yield
was developed in the United States by the Pacific Southwest
Interagency Committee (PSIAC) in 1968. The method is
commonly referred to as the PSIAC method. The method was
developed for use in the southwestern U.S. where the climate
is arid to semiarid, and it is intended for use on drainage
basins with areas larger than 25 km^2. However, it has been
applied to drainage basins as small as 0.05 km^2 with good
results (Shown, 1970).
The PSIAC method (1968) involves the qualitative rating
of a drainage basin based on nine factors that represent
surface geology, soils, climate, runoff, vegetation, land
use, and erosion characteristics. Each of these factors is
assigned a numerical rating based on field examination of the
basin (see Table 1.1). Characteristics of each of the nine
factors that result in high, moderate, or low sediment yield
estimates are shown in Table 1.1. For example, surface
geology would be assigned a value of 10 in an area underlain
by shale or unconsolidated deposits of silt and clay. The
sediment yield rating assigned to a drainage basin is the
sum of values for the appropriate characteristic for each
of the nine factors. Conversion of rating classes to
sediment yield classes is shown in Table 1.2.

Table 1.1. Rating ranges for the factors evaluated in the
           Pacific Southwest Interagency Committee method
           for estimating sediment yields using basin
           characteristics

Factor	Rating range	Main characteristics considered	
Surface geology	0-10	Rock type.	Weathering.
		Hardness.	Fracturing.
Soils	0-10	Texture.	Salinity.
		Aggregation.	Caliche.
		Shrink-swell.	Organic matter.
		Rockiness.	
Climate	0-10	Storm frequency, intensity and duration.	
		Snow.	
		Freeze-thaw.	
Runoff	0-10	Volume per unit area.	
		Peak flow per unit area.	
Topography	0-20	Steepness of upland slopes.	
		Relief.	
		Fan and flood plain development.	
Ground cover	-10-10	Vegetation.	
		Litter.	
		Rocks.	
		Understory development under trees.	
Land use	-10-10	Percentage cultivated.	
		Grazing intensity.	
		Logging.	
		Roads.	
Upland erosion	0-25	Rills and gullies.	
		Landslides.	
		Wind deposits in channels.	
Channel erosion and sediment transport	0-25	Bank and bed erosion.	
		Flow depths.	
		Active headcuts.	
		Channel vegetation.	

Table 1.2. Rating ranges and corresponding estimated sediment yield ranges

Rating class ranges	Estimated annual sediment yield ranges (tonnes/hectare)
>100	>18.3
75-100	6.1-18.3
50- 75	3.0- 6.1
25- 50	1.2- 3.0
0- 25	< 1.2

The PSIAC method was tested in 28 small drainage basins in Wyoming, Colorado, and New Mexico ranging in size from 0.05 to 96 km^2 where sediment yield data has been collected for several years (Shown, 1970). It was found that estimates of sediment yield made with the PSIAC method correlated closely with measured sediment yields. Although the PSIAC method is qualitative and subject to the bias of individual investigators, it is useful for broad planning purposes and extrapolation of existing data to ungauged areas.

A description of all the empirical methods that have been developed and tested in the field for measuring upland erosion and sediment yield is beyond the scope of this book. A study of the published literature generally will provide descriptions of methods and results that can be adapted to a variety of environments and erosion problems. It should be emphasized that empirical methods may offer a solution to an immediate problem, but understanding erosion-sediment transport processes is essential to development of a physically based model.

REFERENCES

Aina, P.O., Lal, R., and Taylor, G.S., 1977, Soil and crop management in relation to soil erosion in the rainforest of western Nigeria, in *Soil erosion: prediction and control*, (Soil Conservation Society of America, Ankeny, Iowa) 75-82.

Brooks, F.L., 1977, Use of the Universal Soil Loss Equation in Hawaii, in *Soil erosion: prediction and control*, Soil Conservation Society of America, Ankeny, Iowa) 23-30.

Campbell, I.A., 1974, Measurements of erosion on badlands surfaces, *Zeitschrift fur Geomorphologie*, Supplementband 21, 122-137.

de Ploey, J. and Gabriels, D., 1980, Measuring soil loss and experimental studies, in *Soil erosion*, ed Kirkby, M.J., and Morgan, R.P.C., (John Wiley & Sons, Ltd., Chichester) 63-108.

Dunne, T., 1977, Evaluation of erosion conditions and trend in *Guidelines for watershed management*, FAO Conservation Guide No. 1, 53-83.

El-Swaify, S.A., and Dangler, E.W., 1977, Erodibilities of selected tropical soils in relation to structural and hydrologic parameters, in *Soil erosion: prediction and control*, (Soil Conservation Society of America, Ankeny, Iowa) 105-114.

Emmett, W.W., 1965, The vigil network-methods of measuremen and a sampling of data collected, *International Association of Scientific Hydrology Publication*, No. 66, 89-106.

Emmett, W.W., and Hadley, R.F., 1968, The vigil network-preservation and access of data, *US Geological Survey Circular*, 460-C.

Evans, W.R., and Kalkanis, G., 1977, Use of the Universal Soil Loss Equation in California. in *Soil erosion: prediction and control*, (Soil Conservation Society of America, Ankeny, Iowa) 31-40.

Hadley, R.F., 1965, Selecting sites for observation of geomorphic and hydrologic processes through time. *International Association of Scientific Hydrology Publication*, No. 66, 217-233.

Hadley, R.F., 1967, Study of slope and fluvial processes, *Revue de Geomorphologic Dynamique*, 17, 158-159.

Hadley, R.F., and Lusby, G.C., 1967, Runoff and hillslope erosion resulting from a high-intensity thunderstorm near Mack, western Colorado, *Water Resources Research*, 3, 139-143.

Hadley, R.F., and Schumm, S.A., 1961, Sediment sources and drainage basin characteristics in upper Cheyenne River basin, *US Geological Survey Water Supply Paper* 1531-B, 137-196.

Holland, M.E., 1969, Colorado State University experimental rainfall-runoff facility, design and testing of rainfall system, *Colorado State University Experimental Station Report*, CER 69-70, MEH 21.

Leopold, L.B., Emmett, W.W., and Myrick, R.M., 1966, Channel and hillslope processes in a semiarid area, New Mexico, *US Geological Survey Professional Paper*, 352-G, 193-253.

Lusby, G.C., 1977, Determination of runoff and sediment yield by rainfall simulation, in *Erosion: research techniques, erodibility, and sediment yield*, ed Toy, T.J., (Geobooks, Ltd., Norwich, England) 19-30.

McCool, D.K., Molnan, M., Papendick, R.I., and Brooks, F.L., 1977, Erosion research in the dryland grain region of the Pacific Northwest: recent developments and needs, in *Soil erosion: prediction and control*, (Soil Conservation Society of America, Ankeny, Iowa) 50-59.

Meyer, L.D., 1965, Symposium on simulation of rainfall for soil erosion research, *Transactions American Society of Agricultural Engineers*, 8, 63-65.

Miller, J.P., and Leopold, L.B., 1963, Simple measurements of morphological changes in river channels and hillslopes, in *Changes of climate, Proceedings of Rome Symposium*, (World Meteorological Organization and UNESCO) 421-427.

Osborne, H.B., Simanton, J.R., and Renard, K.G., 1977, Use of the Universal Soil Loss Equation in the semiarid Southwest, in *Soil erosion: prediction and control*, (Soil Conservation Society of America, Ankeny, Iowa) 41-49.

Pacific Southwest Interagency Committee, 1968, Report on factors affecting sediment yield in the Pacific Southwest area, *Water Management Subcommittee, Sedimentation Task Force Report*.

Roose, E.J., 1977, Use of the Universal Soil Loss Equation to predict erosion in West Africa, in *Soil erosion: prediction and control*, (Soil Conservation Society of America, Ankeny, Iowa) 60-74.

Schumm, S.A., 1967, Erosion measured by stakes, *Revue de Geomorphologie Dynamique*, 4, 161-162.

Shown, L.M., 1970, Evaluation of a method for estimating sediment yield, *US Geological Survey Research, Professional Paper*, 700-B, B245-B249.

Singer, M.J., Huntington, G.L., and Sketchley, H.R., 1977, Erosion prediction on California rangeland: research developments and needs, in *Soil erosion: prediction and control*, (Soil Conservation Society of America, Ankeny, Iowa) 143-151.

Wischmeier, W.H., 1976, Use and misuse of the Universal Soil Loss Equation, *Journal of Soil and Water Conservation* 31, 5-9.

Wischmeier, W.H., Johnson, C.B., and Cross, B.V., 1971, A soil erodibility nomograph for farmland and constructqon sites, *Journal of Soil and Water Conservation*, 26, 189-193.

Wischmeier, W.H., and Smith, D.D., 1958, Rainfall energy and its relationship to soil loss, *Transactions American Geophysical Union*, 39, 285-291.

Wischmeier, W.H., and Smith, D.D., 1965, Predicting rainfall-erosion losses from cropland east of the Rocky Mountains, *US Department of Agriculture, Agricultura Handbook*, No. 282.

Wischmeier, W.H., and Smith, D.D., 1978, Predicting rainfall erosion losses - Guide to conservation planning, *US Department of Agriculture, Agriculture Handbook*, No. 537.

Wolman, M.G., 1977, Changing needs and opportunities in the sediment field, *Water Resources Research*, 13, 49-54.

2.

Soil loss estimation:
a modelling technique

H. A. Elwell

Introduction

The American Universal Soil Loss Equation (USLE)
(Wischemeier and Smith, 1965) has stimulated wide interest
in prediction techniques among conservation conscious
countries. Some have taken the term "universal" literally
and have adopted the American factor values. Others have
seen the dangers in this approach and have put the little
data they have available, towards developing their own
factor values. This has invariably left them without
independent means of assessing the predictive accuracy of
these local factor values. They are, therefore, now faced
with the daunting task of developing an almost inexhaustible
list of correct local factor values as quickly as possible.
This often has to be achieved on a meagre budget and with a
lack of skilled manpower. Consequently they lose interest
in prediction techniques believing that development costs
are will beyond their resources.

Zimbabwe has passed through this phase. The USLE and
its factor values have been tested and found to be
inappropriate under local conditions. Estimates varied
from 50% too low for bare fallow soils to 100% too high for
cropped plots. Furthermore, it has been decided that the
country cannot afford the expensive research programmes
required to build up appropriate local factor values and
many other difficulties and disadvantages were encountered
(Wendelaar, 1978). Nevertheless, an estimation system for
rainfall soil losses was urgently needed and a new method
was developed. This has become known as SLEMSA (Soil Loss
Estimator for Southern Africa).

SLEMSA is a mathematical modelling approach, the purpose
of which is to bring together all sources of information
into a formal arrangement representing the best advice
available. Consequently, established theory, expert opinion
based on field experience, laboratory test data, and results

Senior Conservation Research Engineer, Institute of
Agricultural Engineering, BOX BW 330, Borrowdale, Harare,
Zimbabwe.

from field-plot measurements, are all possible data sources. Useful models can therefore be constructed starting with relatively unsophisticated information, while decisions concerning the application of the model will depend upon the urgency of the problem and the degree of confidence the designer has in its predictions. Because of their low cost a variety of models can be constructed to suit specific field conditions or special purposes. Furthermore they yield useful data at all stages of model building and parts of the model can be selected for improvement in line with priorities and availability of finance.

One such model has been developed to estimate annual soil losses from arable lands on the Zimbabwe highveld. Rill and gully erosion on these lands have been successfully controlled by terraces, called contour ridges in Zimbabwe but severe sheet erosion continues to deplete the soil resources. The model was devised to estimate soil losses from sheet erosion arising from the agricultural practices on the land between the terraces. It has been incorporated into a design procedure the purpose of which is to develop safe farming systems for arable lands.

Safe systems are developed by setting tolerable soil loss levels as target figures which the designer tries to achieve by modifying the slope length, slope percent, tillage treatments, cropping practices and the management standards.

The model has been built from a limited amount of field plot data supplemented by expert opinion from a multidisciplinary team of specialists. At its present stage of development, it will correctly rate field practices in order of erosion hazard and has given good predictions of absolute soil loss for a wide range of common cropping conditions on the Zimbabwe highveld. Its suitability must be tested before it can be safely applied to other localities. However, at the present stage of knowledge it seems likely that, although the control variables will not change, the submodel relationships will vary with relatively discrete changes in climate. Further details can be obtained from the literature (Elwell, 1978).

When fully developed, it is estimated that the prediction technique will have required less than one sixth the capital and one third the labour of that needed to develop the USLE to an equivalent degree of proficiency. And there is every hope that this can be improved upon.

Because of the low cost, the economic use of available expertise and the fact that useful models can be created from even unsophisticated data bases, the technique is considered particularly suited to countries which cannot support expensive research programmes, and in circumstances where a decision-making aid is urgently required to combat soil erosion.

Model building and testing

Model building follows four clearly defined steps:

1. Identify and quantify the major control variables.

2. Propose relationships, called submodels, between carefully selected groups of control variables and soil loss.

3. Formulate a main model expressing the relationships between the submodels.

4. Test the model.

When the form of the relationship between variables is unknown, the simplest linear assumptions are adopted as a starting point. For instance the factors may be assumed to be additive, subtractive, or to interact as simple products. This assessment may be modified later in the light of subsequent research.

Control variables

Soil erosion is the outcome of a large number of causative factors of varying importance, which interact in a complex manner. Consequently the modeller is in danger of including a mass of relatively unimportant data unless he approaches his task logically and takes certain simplifying steps.

The procedure adopted in building the Zimbabwe model was to divide the soil erosion environment into its four physical systems: climate, soil, crop and topography.

The influence of natural properties and man-induced changes were then accounted for within the appropriate system. For instance, natural soil properties, tillage treatments and past and present cropping practices and management standards, all influence the erodibility of a soil and were therefore studied collectively as part of the soil system. Similarly, crop type, planting dates, planting density and management levels were the principal factors considered within the crop physical system. In this way it was possible to identify and quantify the major control variables within each system.

Five control variables have been identified in the Zimbabwe programme: seasonal rainfall energy (E); the amount of rainfall energy intercepted by the crop (i); soil erodibility (F); slope length (L); and slope percent (S).

The control variables are the major overriding factors controlling soil losses within each system and, ideally, they should be easily measurable and rational. They are the "bricks" of model building and, once identified, should remain a consistent feature of a particular main model.

Submodels

In much the same way that many different shapes of structures can be created from the common house brick, a variety of hypotheses (submodels) can be constructed from the control variables to suit specific purposes or field conditions. These hypotheses may be built up from elements of existing theory, laboratory experimental results, soil loss data from field plots, or they may merely reflect the opinion of the relevant expert.

In the case of the highveld model, the control variable were arranged into three carefully thought out submodels: a principal submodel to estimate soil loss from bare soil, a submodel to account for cropping practices, and one to account for differences in topography. The first two submodels were developed from a limited amount of field plot data supplemented by expert opinion; the third one was derived from the slope factor relationship of the American Universal Soil Loss Equation. A correction factor is applied to the topographic submodel, X, to account for crop ridging practices. The submodels are shown in Figs 2.1 to 2.5.

The main model

The main model expresses the relationship between the submodels. The Zimbabwe submodels were formulated to interact as simple products in the main model as follows:

$$Z = K . X . C$$

where Z = the predicted mean annual soil loss, t/ha/year, from the land under evaluation.

K = the mean annual soil loss, t/ha/year, from a standard conventionally-tilled field plot - 30 m x 10 m at a 4.5% slope, for a soil of known erodibility, F, under a weed-free bare fallow.

X = the ratio of soil loss from a field slope of length L m and slope percent S, to that lost from the standard plot.

C = the ratio of soil loss from a cropped plot to that from bare fallow.

Model testing

A soil loss estimation model can be tested at three levels depending on the data available.

Firstly, it can be judged purely on an observational level as to whether it confirms field experience. This is a qualitative test to ensure that the model predicts high and low soil losses in circumstances where this is observed to be the case. The most sophisticated level it is possible to achieve with this kind of test is to see whether the estimation technique correctly rates practices in order of their erosion hazard. When testing the model in this way it is essential to visit specific sites of interest so that the details of the site and farming practices can be recorded.

18

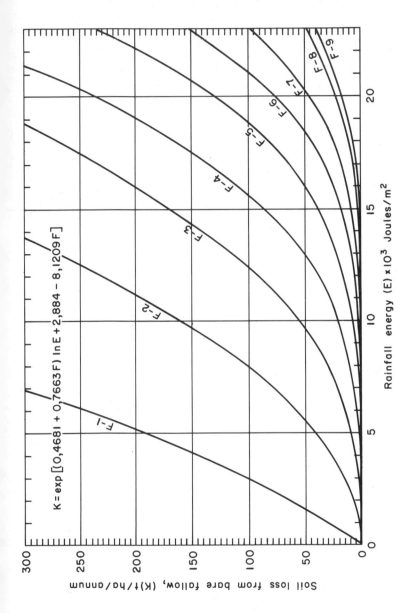

Fig 2.1. Relationships between K, E, and soil erodibility.

$$K = \exp[(0,4681 + 0,7663F) \ln E + 2,884 - 8,1209F]$$

Soil loss from bare fallow, (K)t/ha/annum

Rainfall energy (E) × 10³ Joules/m²

19

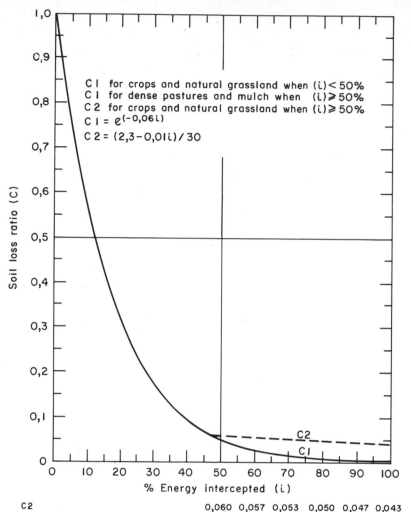

C I for crops and natural grassland when $(i) < 50\%$
C I for dense pastures and mulch when $(i) \geqslant 50\%$
C 2 for crops and natural grassland when $(i) \geqslant 50\%$
$C1 = e^{(-0,06i)}$
$C2 = (2,3 - 0,01i)/30$

| C2 | | | | | | 0,060 | 0,057 | 0,053 | 0,050 | 0,047 | 0,043 |
| CI | 1,0 | 0,55 | 0,30 | 0,17 | 0,09 | 0,050 | 0,029 | 0,015 | 0,008 | 0,005 | 0,002 |

Fig 2.2. Crop cover model.

Supposition or memory should not be relied upon as factors
from different situations are often confused with one another
     Secondly, the model should be a good learning tool.   It
should not only confirm field experience but should also
explain why certain treatments give high soil losses whereas
others do not.   It is particularly valuable in this context
when it gives rational reasons for previously unexplained
phenomena and reveals misconceptions and wrong assumptions
in conservation thinking and planning.
     Thirdly, the predictive accuracy of the model can be
tested against actual soil loss data from field plots.   It
is essential, however, that these data should not previously
have been used to develop the submodels.   This independence
of the data is important for the test to be valid.

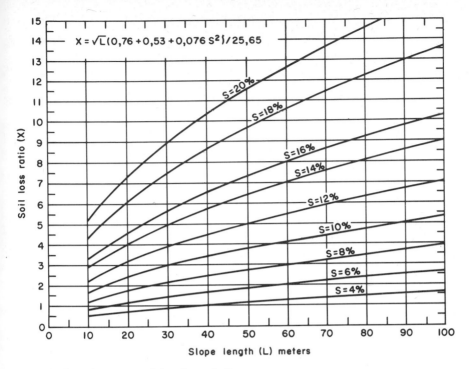

$$X = \sqrt{L}(0,76 + 0,53 + 0,076\ S^2)/25,65$$

Fig 2.3. Topographical model.

   The first two tests are appropriate to models which give
comparative rather than absolute values of soil loss; for
instance, those developed from expert opinion. The third
test is more suitable for models required to give absolute
values such as those developed from field plot data. However,
all possible data sources are often exploited when building
a model and the resultant model is then tested at all three
levels. This was the case with the Zimbabwe model.

Model development and testing is a continuous process. So
far the Zimbabwe model has performed well in comparative tests
and has proven to be a good learning tool. In addition it
has given useful predictions of absolute values of soil loss.
In these tests, the model predictions were compared to actual
measured mean annual soil losses from independent field plots.
Soil losses were estimated to within 1.7, 2.6, and 4.3 t/ha/
year with 50%, 70% and 90% confidence respectively. The model
can also be tested against a single year's soil loss data
provided that the values of the control variables are accurately measured on the site. This latter point is important
because the model can be validated in one year in contrast to
the thirty years claimed necessary for the USLE (Soil
Conservation Society of America, 1976).

Design

Data sources
   Models are developed from 'fundalmental data'. However,
before they can be applied to field problems, it is necessary
to know the value of the control variable at the design site.
Therefore banks of local 'user data' information have to be

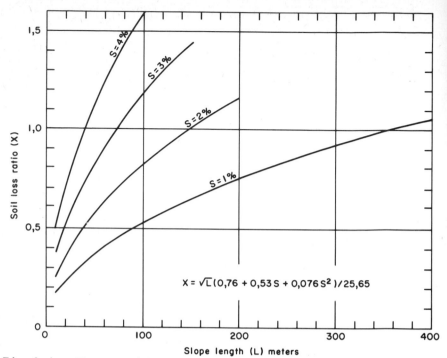

$$X = \sqrt{L}\,(0,76 + 0,53\,S + 0,076\,S^2)/25,65$$

Fig 2.4.   Topographical model for S = 1% to 4%.

compiled giving the variation in values of the control
variables over the geographic area for which the model is
required.   Examples of typical user data information are
given in Fig 2.6. and Tables 2.1. to 2.6.

Procedure

The steps in the design procedure are as follows:

i)   Obtain the mean annual rainfall for the area from
published meteorological reports and note whether
the site is in a guti or non-guti rainfall area.
The term guti refers to locations experiencing
significant amounts of low intensity rainfall.
These are restricted to the south and east of the
country.   Estimate E using Fig 2.6.

ii)   From Table 2.1. assess the basic erodibility value
of the soil, Fb, under conventional tillage.

iii)   From Fb and Table 2.2. assess the soil erodibility
value, Fm, applicable to the soil treatment for
each year in the rotation.

iv)   Enter Fig 2.1. with Fm and E and read off the value
of K for each year in the rotation.

v)   From details of crop yields and planting dates, read
off the appropriate value of i for each crop.   Tables
2.4. to 2.6.

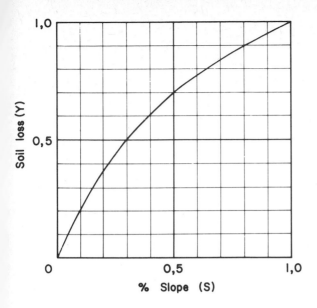

Grade	% Slope
I in 400	0,25
I in 300	0,35
I in 250	0,40
I in 200	0,50
I in 150	0,67
I in 100	1,00

Equation

$$Y = S/(0,572S + 0,428)$$

Where Y is a ratio of X
for S = 1% and S is the
grade of the crop ridges.

Fig 2.5. Crop ridge grade model.

vi) Enter the i values into Fig 2.2. and obtain C for each year.

vii) For unridged lands, L is the distance between terraces and S is the steepest slope of the land. Using values of L and S measure on the site, read off X from Figs 2.3. or 2.4. as appropriate.

viii) Where crop ridge grades are greater than 1%, the value of S is the maximum grade anticipated on the crop ridges in any part of the land, and L is the length of these same ridges. In this case the value of X is read from Figs 2.3. or 2.4. as before

ix) Where crop ridges are less than 1% grade, the value of X is read from Fig 2.4. using S = 1% and L is taken as the maximum length of the crop ridges measured in the land. This value of X is then corrected by multiplying it by the ratio of Y obtained by putting the actual crop ridge grade into Fig 2.5.

x) Determine the mean annual soil loss for the rotation by summing the soil losses for the individual years and dividing by the number of years in the rotation.

xi) Select a suitable target level of soil loss from Table 2.3. and compare with the estimated mean for the rotation.

xii) When the target level is exceeded, select and test alternative systems.

xiii) Choose the most practical system in consultation with the farmer.

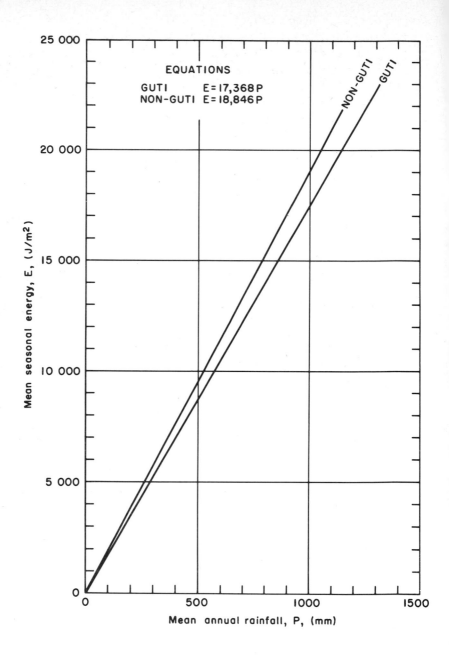

Fig 2.6. Selection of E from mean annual rainfall

Example (Tobacco Rotation)

a) Climate   : mean annual rainfall $E = 16000 \ J/m^2$
              850 mm (non-guti)

b) Soils   : well drained sands $Fb = 4$ (Table 2.1.)
              (5G)

c) Terraces : land at 6% slope $S = 6\%$

              at 24 m spacing $L = 24 \ m$

d) Crops and Tillage:

     1st year (Tobacco)

     - planted out 1st. Nov. $i = 40\%$ (Table 2.4.)
       yielding 2000 Kg/ha

     - Large crop ridges at
       6% grades $Fm = 3$ (Table 2.2.)
       (Fb minus 1 for steep ridges)

     2nd year (Maize)

     - emerging 15th. Nov. yielding $i = 45\%$ (Table 2.4.)
       6000 Kg/ha

     - Conventional tillage $Fm = 3$ (Table 2.2)
       (fine tilth; Fb minus =
       for soil loss from tobacco)

     3rd year (Katambora Rhodes pasture)

     - planted in early Nov. $i = 96\%$ (Table 2.6.)
       under good management
       (Fb minus 0,5 for soil
       losses from maize)

     4th and 5th year (Katambora Rhodes pasture)

     - lightly grazed in summer $i = 90\%$ (Table 2.6.)
       and winter
       (Fb plus 1 for second year $Fm = 5$ (Table 2.2.)
       fallow)
       (Fb plus 2 for third year $Fm = 6$ (Table 2.2.)
       fallow)

e) Target level:

     - design target level of $Zt = 5$ (Table 3.)
       soil loss (t/ha/year)

Table 2.1.   Selecting Fb

| | | Texture of top soil |||
Soil Group	Soil Family	A,X	B,C	D,E,F,G
Regosol	1K	4		
Lithosol	2	2	2,5	4
Vertisol	3B, 3E, 3X			5
	3S			4,5
Siallitic	4E		3,5	<u>4</u>
	4X			4
	4S			3,5
	4P, 4M	<u>3,5</u>	4	
	4G	3		
Fersiallitic	5E		5,5	<u>6,0</u>
	5X			6,0
	5S		4	<u>4</u>
	5G	<u>4</u>	5	
	5P	<u>4</u>	4,5	
	5M	4,5		
	5F	4	<u>5</u>	
	5A	3,5	<u>4</u>	
Paraferrallitic	6G	<u>4,5</u>	5	
Orthoferrallitic	7E			7
	7S			5,5
	7G	5,5	<u>6,5</u>	6,5
	7M	5	<u>6</u>	
Sodic	8N	<u>1</u>	1,5	1,5
	8n, 8h	<u>1</u>	1,5	2

N.B.   Where 2 or more values are given, the most
commonly occurring soil is underlined.

26

Table 2.1.(continued)

Key to symbols:

Soil Group		Soil Family	
1.	Regosol	A	Arenaceous
2.	Lithosol	B	Basalt
3.	Vertisol	C	Colluvium
4.	Siallitic	E	Basic rocks
5.	Fersiallitic	F	Micaceous rocks
6.	Paraferrallitic	G	Granites and coarse gneisses
7.	Orthoferrallitic	h	Saline-sodic
8.	Sodic	K	Kalahari sand
		M	Sandstones and quartsites
Texture		n	Weakly sodic
A	Sand	N	Strongly sodic
X	Loamy sand	P	Fine-grained siliceous gneisses
B	Sandy loam	S	Argillaceous sediments
C	Sandy clay loam	U	Alluviums
D	Clay loams	X	Ultra basic rocks
E	Sandy clay		
F	Clay		
G	Heavy clay (vertisol)		

Table 2.2. Derivation of Fm from Fb

(Fm = Fb $\pm$ the sum of the correction factors
given below obtained from A + B + F or
A + C or D or E as appropriate)

Practice	Factor
A. Soil losses from the previous year	
- Less than 10 t/ha	0
- 10 to 20 t/ha	-0,5
- Greater than 20 t/ha	-1,0
B. Ridging practices	
B1 Crops on big ridges (not less than 200 mm high consolidated)	
- Flatter than 1% grades with tie-ridges	+1,5
- Flatter than 1% grades without tie-ridges	+1,0
- Between 1 and 2% grades	0
- Over 2% grades	-1,0
B2 Crops on small ridges (less than 200 mm high unconsolidated)	
(The construction of undersized ridges is not advocated!)	
- Flatter than 1% grades	-1,0
- Between 1 and 2% grades	0
- Over 2% grades	-1,0
C. Annual crops planted on the flat	
C1 Planting and ploughing directions	
- Operations "level" or "on-contour"	0
- Operations at angles to the contour ridges	-0,5
C2 Tillage techniques	
- Plough (250 mm), roll and disc harrow to give a fine tilth, e.g. conventional tillage	0
- Plough (250 mm) and roll to give a rough tilth e.g. rough-conventional	+0,5
- Ripped to 300 mm and lightly disced to 80 mm, e.g. rip and disc	0
- Plough (250 mm), plant in strips of tilth made by the tractor wheels, leaving clods in the inter rows, e.g. wheel-track planting	+1,0

Table 2.2. (continued)

- No ploughing or discing with crop
    tine-planted   -0,5
    e.g. zero tillage
- Very fine powdery tilth, e.g. cotton tilth   -0,5
- Rip to 300 mm, cultivate and ro ,
    e.g. con-till   0

D. Fallows and leys (good management)
    - 1st year fallow or ley   0
    - 2nd year fallow or ley   +1,0
    - 3rd year and subsequent years   +2,0
    - Permanent pastures and veld in good
        condition   +2,0

E. Perennial crops and orchards
    - No tillage and cultivated mechanically   -0,5
    - No tillage and herbicde weed control   0
    - No tillage and soil showing a marked
        improvement in structure, such as
        occurs under heavy mulch   +2,0

F. Irrigation
    - Light textured irrigated soils (sand and
        loamy sand topsoil)   -0,5
    - Heavily fertilised and irrigated pastures   +3,0

Notes All practices are judged by comparing them to
      conventional tillage. If a practice either increases
      run-off, or reduces the soil's resistance to being
      broken down by raindrop action, it is given a negativ
      rating. Conversely, practices which store water and
      resist detachment are given a positive rating.

      The designer can interpolate between values if he is
      confident in his ability to do so.

Table 2.3. Target levels of soil loss

Texture	Target level Z_t (t/ha/year)
Sand	5
Loamy sand	5
Sandy loam	5
Sandy clay loam	4
Clay loam	4
Sandy clay	4
Clay	3
Heavy clay	3

Notes   Ideally, the rate of soil loss should be related to the rate of soil formation. However, formation rates are thought to be of the order of 1 t/ha/year or less. This level of soil loss is too low to be a practical target level.
   The value set for target level should be possible to achieve when good conservation practices are applied. Therefore, one way of setting a target level is to adopt the value of soil loss predicted by the model for an acceptable farming system.

f) Calculations:

Crop		K (Fig 2.1.) (t/ha/year)	C (Fig 2.2.) (ratio)	X (Fig 2.3.) (ratio)	Z = KCX (t/ha/year)
Tobacco	(T)	200	0,09 (Curve C2)	1.25	22.5
Maize	(M)	200	0.07 (Curve C2)	1.25	17.5
Grass	(G)	150	0.045 (Curve C2)	1.25	8.4
Grass	(G)	50	0.005 (Curve C1)	1.25	0.3
Grass	(G)	25	0.005 (Curve C1)	1.25	0.2
					48.9

Mean soil loss for the rotation = 9.8 t/ha/year

Target level = 4   t/ha/year

g) Decision:

Investigate alternative protective systems!

h) Alternatives:

A. An extra year of grass added to the rotation (TMGGGG)
   - Z = 0.2 in sixth year of rotation.

B. Maize removed and an extra year of grass added (TGGGG)
   - 1st year pasture can now be planted in April (Curve C1).

C. Tobacco grown on big ridges at 1 in 200 grades (TMGGG)
   - Fm becomes 5 for tobacco and rises to 4 for maize and first year of pasture.
   - L = 200 m and S = 1 in 200 for tobacco, giving a new value X = 0.533.

D. Tobacco as per alternative C for two years (TTGGG)
   - Maize deleted.
   - Fm for grass becomes 4 and it is planted in April (CURVE C1).

i) Summary of soil losses:

Alternatives	T	M	T	G	G	G	G	Mean
As farmed	22.5	17.5		8.4	0.3	0.2		9.8
A	22.5	17.5		8.4	0.3	0.2	0.2	8.1
B	22.5			0.8	0.3	0.2	0.2	4.8
C	2.4	9.6		6.2	0.3	0.2		3.7
D	2.4		2.4	0.4	0.3	0.2		1.1

j) Decision:
    The farmer should be persuaded to change to alternative C or D. Alternative D is clearly more profitable, and preferable from a soil conservation point of view; however, specialist advice should be sought with respect to nematode control

TABLE 2.4. i Values for crops
(Use Cover Model C2)

Crop	Yield	i% For emergence dates										
	kg/ha	1/S	15/S	1/O	15/O	1/N	15/N	1/D	15/D	1/J	15/J	1/F
Cotton	500	43	41	39	34	29	23	16	11	7	3	1
	1 000	62	59	55	49	41	32	24	16	9	5	2
	1 500	72	69	65	57	48	38	28	19	11	6	2
	2 500	84	79	75	66	56	44	32	22	13	6	3
	3 500	92	87	82	72	61	48	35	24	14	7	3
	4 500	95	89	84	74	63	49	36	25	14	7	3
Cowpeas	t/ha											
Hay	0.5	35	38	41	44	44	42	36	30	23	22	9
	1.5	54	58	62	66	67	63	55	46	35	24	14
	>2.5	65	71	76	80	81	77	67	56	43	30	17
Silage	2	26	27	29	31	32	31	29	25	20	14	8
	6	43	46	50	53	54	53	49	42	34	24	14
	>10	61	64	70	74	83	74	69	59	48	37	21
Green Crop	2	2	4	6	8	9	10	10	9	8	5	3
	6	5	10	17	22	26	30	30	26	22	15	9
	>10	7	14	24	31	35	41	41	36	31	21	12
Groundnuts	kg/ha											
(unshelled)	250	32	35	37	35	32	26	21	15	10	6	3
	750	50	55	57	55	49	41	32	24	16	9	5
	1 500	64	70	73	70	62	52	41	30	20	12	6
	3 000	71	77	80	77	69	58	45	33	22	13	6
Maize	500	14	16	18	18	17	15	12	10	7	4	2
	1 000	18	21	23	23	22	20	16	13	9	6	3
	2 000	24	27	29	29	28	25	21	16	12	7	4
	4 000	34	38	42	42	40	36	29	23	17	10	5
	6 000	43	48	52	53	51	45	37	29	21	13	6
	8 000	50	56	61	62	59	53	43	34	24	15	8
	10 000	55	63	68	69	66	59	48	38	27	17	8
Rice	500	31	33	33	33	31	28	23	19	14	9	5
	1 500	54	56	56	56	53	47	40	32	24	16	9
	2 500	69	71	72	71	67	60	51	41	31	20	11
	3 500	79	81	82	81	77	69	58	46	35	23	13
	4 500	84	87	88	87	82	74	62	50	37	25	13
Sorghum	500	32	33	33	32	29	25	21	16	10	6	3
	1 000	39	40	40	38	35	30	25	19	12	7	3
	2 000	47	49	49	47	43	37	30	23	15	9	4
	4 000	72	74	74	71	65	55	46	35	23	13	6
	6 000	81	84	84	80	73	63	52	40	26	15	7
Soyabeans	250	22	24	25	25	23	21	17	13	9	5	3
	750	34	37	38	38	36	32	27	20	14	8	4
	1 500	51	54	57	57	54	48	39	29	20	12	6
	2 500	69	74	77	77	73	64	53	40	28	16	8
	4 000	80	86	90	90	85	76	62	47	32	19	10

TABLE 2.4. (continued)

Crop	Yield kg/ha	i% For emergence dates										
		1/S	15/S	1/O	15/O	1/N	15/N	1/D	15/D	1/J	15/J	1/F
Sunflowers	125	5	7	8	10	11	10	9	8	5	3	2
	375	11	16	19	22	24	23	21	17	12	7	3
	750	20	28	34	39	43	42	38	31	21	12	6
	1 250	28	41	50	57	62	61	55	45	30	18	9
	1 500	31	44	53	61	67	65	59	48	32	19	10
Tobacco (flue cured) above 1100 m altitude	500	4	6	7	8	9	8	7	6	4	2	1
	1 000	8	12	16	18	19	18	16	13	8	5	2
	1 500	12	18	24	27	30	28	25	20	13	7	3
	2 000	17	25	32	36	40	38	34	27	17	9	4
	2 500	21	31	40	46	50	47	42	34	22	12	5
	3 000	25	38	49	55	61	58	51	41	27	14	7
Tobacco (flue cured) below 1100 m altitude	500	3	5	5	6	7	6	6	5	3	2	1
	1 000	6	9	13	14	15	14	13	10	6	4	2
	1 500	9	14	19	21	24	22	20	16	10	5	2
	2 000	13	20	25	28	31	30	27	21	13	7	3
	2 500	16	24	31	36	39	37	33	27	17	9	4
	3 000	20	30	38	43	48	46	40	32	21	11	5
Velvet beans Hay	t/ha 0.5	56	58	57	54	47	40	32	23	15	9	4
	2.0	81	83	83	78	69	57	46	33	21	13	6
	3.5	94	98	96	91	80	67	54	39	25	15	7
Silage	1.5	27	29	30	29	27	23	19	14	9	5	2
	4.5	52	55	57	56	51	45	36	26	16	10	4
	7.5	65	69	71	70	64	56	45	32	21	13	5
	10.5	72	76	79	78	71	62	50	36	23	14	6
Green Crop	1.5	12	15	18	20	19	17	14	10	6	4	2
	4.5	20	25	30	33	32	29	24	17	11	7	3
	7.5	32	40	48	53	51	46	38	27	17	11	5
	10.5	37	46	55	61	59	53	43	31	20	12	5
Weed Fallow	(High Ferti -lity only)	93% for most probable starting date of the rains										

Note: 1/S; 1/O = 1st September, October etc.
15/S; 15/O = 15th September, October etc.

TABLE 2.5. i Values for crops planted into mulch
(Total i = i for mulch alone + a proportion of i for crop alone)

Treatment	Examples	Total i %	Cover Curve
Maize stover slashed and left on the surface	Zero till Con till	50% + 0.45 i	C1
Maize stover disced in to 100 mm	Rip and disc	30% + 0.7 i	C2
Maize stover ploughed in	Conventional tillage	i	C2
Stover burned off		i	C2
Planting direct into wheat stover	Zero till	75% + 0.25 i	C1

Note: Maize yields of above 5000 kg/ha and wheat yields of above 2.5 t/ha are assumed.

TABLE 2.6. i Values for grass pastures in a tobacco rotation
(well managed with potentially 100% cover but limited nitrogen)

Grazing pressure	Year in the rotation	i% For emergence dates										Appropriate Cover Curve
		1/O	15/O	1/N	15/N	1/D	15/D	1/J	15/J	1/F	15/F	
Light summer and winter grazing.	1st yr	96	94	89	80	68	55	42	29	16	8	* C2
	2nd yr	90										** C1
	3rd yr etc.	90										** C1
Heavy winter grazing. Light summer grazing.	1st yr	as for 1st year above										* C2
	2nd yr	70										** C1
	3rd yr etc.	70										** C1

Notes: It is assumed that the pasture is not grazed during the 1st year. If
* planted in April use curve C1 and curve C2 for November planting.
** Use cover curve C2 for 2nd and 3rd years of the ley when pastures
are heavily grazed during the summer months.

Conclusions

The modelling approach has many advantages over conventional techniques. The principal ones being: the maximum use of limited resources, including finance, facilities and skill; and the provision of an early basis for decision making. However the limitations of models must be fully understood.

The accuracy with which a model can predict the real-life situation depends upon the reliability of its data base and how well the model is constructed. However, even the most well-conceived model developed from field data will not give perfect answers. The most that can be expected is an answer which is of the right order of magnitude and has a good chance of being correct. This is because of the uncontrolled variations occurring on agricultural lands - a feature which bedevils all predictive techniques.

A model is more reliable in the middle ranges of the input variables than at the extremes, and can be applied with confidence only to those conditions for which its predictive capability has been established.

Crude models should not be used for fine adjustments. Thus, a model developed from unsophisticated data may satisfactorily distinguish between practices yielding high and low soil losses, but it is unlikely to correctly discriminate between practices which give similar soil loss values.

Nevertheless, crude empirical models such as those based on expert opinion have much to offer in that the novice now has easy access to the most experienced advice available. As more research data becomes available and the control variables begin to be identified, models become less empirical and more rational. This is clearly a move in the right direction. Models which give rational explanations of observed phenomena will be of interest to hydrologists in their search for methods of assessing the potential sediment yield from catchments.

The Zimbabwe highveld model is in a relatively early stage of development, the weakest link being the bare fallow submodel K. Steps are in hand to improve the submodel and provide a better quantitative definition of the soil erodibility, F, based on research data. However, the model has performed sufficiently well in comparative field tests, and as a predictor of soil losses from independent field plots, to have application in design, planning and extension.

REFERENCES

Elwell, H.A., 1978, Compiled Report of the Multidisciplinary Team on Soil Loss Estimation, *Departmental Report, Department of Conservation and Extension*, Causeway, Zimbabwe.

Soil Conservation Society of America, 1976, *Soil erosion: Prediction and Control*, (Soil Conservation Society of America, Ankeny Iowa).

Wendelaar, F.E., 1978, Applying the Universal Soil Loss Equation in Rhodesia, *Departmental Report, Department of Conservation and Extension*, Causeway, Zimbabwe.

Wischmeier, W.H., and Smith, D.D., 1965, Predicting rainfall erosion losses from cropland and east of the Rocky Mountains, *US Department of Agriculture, Agriculture Handbook*, No. 282.

3.

Measurement of sediment yields

P. R. B. Ward

The standard approach used by most government agencies for measuring sediment yield is to establish permanent stations at river sites where the amount of sediment passing that station (unit weight per day) is measured (a) periodically, or (b) continuously.
Methods of type (a), suitable for determining mean values averaged across the cross-section, are carried out by lowering instruments to the required measuring point and sampling. The results are totalled to determine a cross-sectional average transport at the time of measurement. These methods do not offer continuous records of sediment transport versus time. Techniques of measuring sediment transport continuously (type b) rely on the operation of a device that monitors suspended sediment concentrations, usually at a single point near the bank. Methods of this type are used advantageously in rivers whose sediment transport characteristics change very quickly from hour to hour. These methods (type b) only sample correctly the part of the sediment load that is fine enough to be well mixed across the section.
An alternative direct approach to measuring sediment yields is to monitor the accumulation of sediment deposits on the beds of reservoirs and at delta fronts where rivers enter the ocean. These measurements are usually carried out seasonally and are thus only suitable for determining annual totals. Amounts of sediment are measured as volumes and accompanying measurements of density must be made if accumulated weights of sediment are to be found. The finest fraction of the sediment is not measured by this method as this fraction remains in suspension and either passes through the reservoir and out over the spillway or, in the case of coastal deltas, passes out to sea. (see chapters 7 and 8).

RIVER GAUGING STATIONS

Rivers transport sediment of a very wide variety of sizes. The finest particles, colloids, clays and silts, are transported in suspension. The coarse particles (sand and gravel) are transported partly in suspension and partly as

Dr Peter R B Ward, Peter Ward & associates Ltd, 3408 West 40th Avenue, Vancouver, B.C., Canada V6N 3B6.

Table 3.1.   Size classification of river sediment (after Lane *et al.*, 1947)

Class name	Size range	
	mm	micron
Boulders	256-4100	
Cobbles	64-256	
Gravel	2-64	
Very coarse sand	1-2	
Coarse sand	0.5-1.0	
Medium sand	0.25-0.5	250-500
Fine sand	0.125-0.25	125-250
Very fine sand	0.062-0.125	62-125
Coarse silt	0.031-0.062	31-62
Medium silt	0.016-0.031	16-31
Fine silt	0.008-0.016	8-16
Very fine silt	0.004-0.008	4-8
Coarse clay size	0.002-0.004	2-4
Medium clay size	0.001-0.002	1-2
Fine clay size	0.0005-0.001	0.5-1
Very fine clay size	0.00024-0.0005	0.24-0.5

bed load by rolling and sliding of individual particles near the bed. Very coarse particles, such as boulders, move only as a bed load. Table 3.1. gives a definition of the various parts of the sediment load, by size, following Lane *et al.* (1947).

The method used to gather information on sediment transpor depends on what the data are to be used for. Prediction of channel bed erosion (downstream of dams for example) requires information on the rate of erosion of bed material and thus measurements of bed material load must be made. Predictions of rates of reservoir siltation depend on knowledge of the total load carried by the rivers supplying the reservoir. Since in most parts of the world the biggest fraction of the total load is sediment moving in suspension, the emphasis in this application should be on suspended load measurements.

Bed material load is a term used by Einstein (1950) and others to describe the part of the sediment load which can be calculated from the hydraulic characteristics of the river flow. Einstein (1964) also proposed that the total sediment load is divisible into two parts. One part, the wash load, moves always in suspension. Its transport rate past a measuring point is limited by the ability of the catchment to supply the material. The river is always able to transport all the wash load supplied to it. The second part of the load the bed material load, consists of the coarser fraction.

Transport of bed material load past a measuring point is
limited by the ability of the flowing water to convey the
load. The bed material load may thus be predicted if the
hydraulic characteristics of the channel are known. The bed
material load moves partly as bed load and partly in
suspension.

The various particle sizes in the bed material move at
different transport rates and thus the size distribution for
bed material (sampled for example from the channel bed at
times of low flow) is very different from the size distribution
of the bed material load.

As an approximate guide Einstein (1950) uses a size equal
to the particle size for which 10% of the bed material is
finer to separate sediment that moves as wash load from
sediment that moves as bed material load. Einstein's
calculation allows the part of the bed material load moving
strictly as bed load to be separated from the part of the bed
material load moving in suspension.

The division between bed load and suspended load is clearly
not a firm one. Most rivers, even sand bed rivers, have
occasional rock bars or constricted rocky sections where
turbulence causes the bed load to go into suspension. One of
the best methods of making measurements of total load in the
field is to find a naturally occuring rocky constriction, or
to construct a turbulence flume (Middle Loup River, Nebraska
(Serr, 1951)) and to measure all of the sediment load using
suspended load sampling in the turbulent reach.

As a first estimate Einstein (1964) gives a value between
10% and 20% as the percentage of the total sediment load which
moves as bed material load. Lane and Borland (1951) give a
table by Maddock which shows that in most categories of rivers
the bed load (defined as that part of the load not measured by
suspended sediment samples) is 2% to 26% of the total load.
Only in one special category (rivers with small suspended
sediment concentrations with sand beds) does Maddock's table
show the bed load to be a large percentage of the total.

Church (1972) describes sediment load measurements on
rivers in an unusual morphogenetic region (the Canadian arctic)
where the bed load constitutes 80% to 95% of the total load.

In designing a sediment sampling programme for engineering
purposes there is a need to obtain reliable approximate
measurements but seldom a need for great accuracy. Indeed,
great accuracy is usually not attainable because the water
discharges themselves (included as part of the suspended
sediment load calculation) are only reliable to ± 3% under
optimum conditions (Carter and Anderson, 1963) and probably no
better than ± 15% during major floods. The errors to be
expected from sampling by very simple inexpensive methods, not
normally in use by Government departments in the United States
and Canada, are discussed in the next section. In certain
circumstances these errors will be seen to be sufficiently
small to warrant consideration of a simple sampling technique,
such as surface sampling with a bucket, as an operational
method. These simple field procedures together with new,
simple methods of laboratory analysis achieve very great
economic savings in the operation of sediment load measurements.

They are particularly valuable for use in developing countries, where it is of utmost importance to ensure that the experimental procedures are technically simple and not prone to operator error.

Considerable care should be given to the selection of a suitable sampling strategy if inefficiencies are to be avoided in routine sampling programmes. Flood flows must be sampled more comprehensively than flows during the rest of the hydrologic year. In many parts of the world a large proportion of the annual runoff is carried during floods. Sediment transport is even more intermittent than water discharge. In arid and semiarid regions, in the dry tropics, and in mediterranean climates, the sediment discharge is so intermittent that it is possible to operate sediment sampling measurements for a few months only and to close stations for the rest of the year. The main aim of the program should be to sample thoroughly the few floods that are significant. The frequency of sampling during these floods depends on the duration of the unit hydrograph for the catchment and is probably best found either by inspecting the flow records or by trial during the first season of operation. At times between floods sampling can be conducted on a daily basis, and once the flow decreases to a certain value (equal to about 3 times the mean annual runoff discharge in the dry tropics) sampling can be terminated altogether. Special equipment has been designed (Water Resources Council Sedimentation Committee 1976, p.37) for operating pumping samplers at various frequencies, depending on flood stage.

Errors incurred in suspended sediment sampling

The transport of suspended sediment past a measuring section fluctuates in time and in space (Vanoni, 1975, p. 321 due to changes in both the water discharge and the suspended sediment concentration. Turbulent fluctuations in the water flow range in size from the smallest (of the order of a fraction of a millimeter) to those of length scale about equal to the channel width and larger, and these fluctuations are associated with changes in suspended sediment concentrations, particularly concentrations of the sand sized fraction. In addition systematic spatial variations of the suspended sediment concentration of the coarser sizes may be expected due to secondary currents at bends.

For these reasons a suitable sediment sampling site for permanent use must be carefully chosen. Einstein (1969) suggests looking for sections where it is clear that the main thread of the flow changes frequently from one side of the river to the other (e.g. permanent crossings, the end of bends ...) so as to obviate the effects of very long (of the order of years) changes of concentration patterns.

No theories are available at the moment to analyse the effects of using only one or a very small number of sampling points across the breadth of the river. It is, however, possible to analyse the errors involved in using one sample to represent the concentration in the vertical using theories on the vertical distribution of velocity and concentration.

40

This has been done by Brooks (1965) for sampling at the mid-depth point and will be extended here to other cases. The discharge of suspended sediment per unit width of stream is given by:

$$g_{ss} = \int_{y_o}^{d} CU \, dy \qquad (3.1)$$

in which C(y) is the sediment concentration at distance y above the bed.

U(y) is the flow velocity at distance y above the bed

d is the depth of flow

y_o is a small value of y taken as the lower limit of integration

C(y) is given by Rouse's distribution (Rouse, 1937) and U(y) by the commonly used logarithmic distribution. The equations are, respectively:

$$\frac{C}{C_a} = \left[\frac{d-y}{y} \cdot \frac{a}{d-a} \right]^{z} \qquad (3.2)$$

$$U = V + \frac{U_*}{k} (1 + \ln y/d) \qquad (3.3)$$

in which C_a is the concentration of sediment at level y = a, $\bar{z}$ is the dimensionless settling velocity ($\bar{z} = w/k \, U_*$), w is the settling velocity of the sediment at the temperature of the river water (see Figure 3.2.), U_* is the shear velocity of the flow, k is von Karmen's constant and V is the depth average velocity. The shear velocity U_* may be found from the mean velocity V using the relationship (see Henderson, 1966).

$$\frac{U_*}{V} = (f/8)^{\frac{1}{2}} \qquad (3.4)$$

where f is the friction factor. As an approximate guide, f = 0.01 to 0.02 for unsmoothed concrete and f = 0.05 to 0.1 for earth channels in rough condition. (Note that the friction factor f used here is that used in United States, not British, practice).

The suspended sediment discharge concentration is (g_{ss}/q), where q is the water discharge per unit width of channel. The ratio (g_{ss}/qC_a) between the suspended sediment discharge concentration and the concentration C_a at the point of measurement is obtained from equations 3.1, 3.2 and 3.3 as

$$\frac{g_{ss}}{qC_a} = (1 + \frac{U_*}{Vk}) \, A^{z} \int_{y_o}^{d} \left[\frac{d-y}{y} \right]^{z} dy + \frac{U_*}{Vk} A^{z} \int_{y_o}^{d} \left[\frac{d-y}{y} \right]^{z} \ln \frac{y}{d} \, dy \qquad (3.5)$$

in which A is the relative sampling height A = a/(d-a). Values of q C_a/g_{ss} as a function of $\bar{z}$ have been evaluated by numerical integration and are shown in Figure 3.1. for several values of a/d. In all cases U_*/V is equal to 0.1 (f = 0.08, a value to be expected for rivers in flood) and k equal to 0.4, giving U_*/Vk equal to 0.25. The lower limit of integration y_o

41

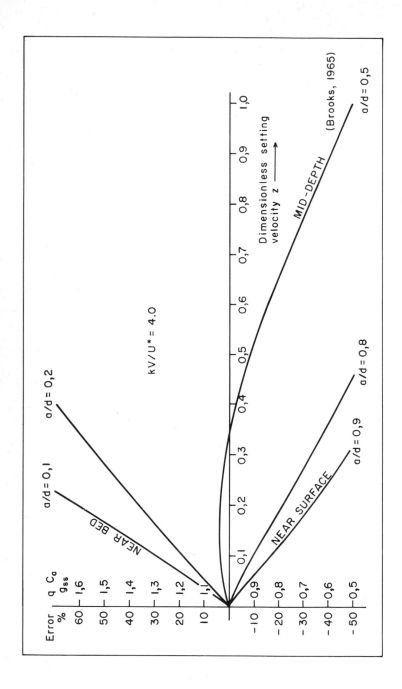

Fig 3.1. Single point sampling error versus dimensionless settling velocity z.

was taken as the level such that $U_* = 0$, giving $y_0/d =$
exp - $(kV/U_* + 1)$. qC_a/g_{ss} is a measure of the error
included by sampling at point $y = a$ instead of using the
(true) suspended sediment discharge concentration (g_{ss}/q)
to determine the suspended sediment load. As qCa/g_{ss} becomes
closer to 1.0, so the sampling error gets smaller.
The use of a single sample taken from near the surface to
represent the suspended sediment discharge concentration is
obviously an approximation, because the sediment concentrations
near the bed are larger that those near the surface.

The approximation becomes less severe for the smaller sizes
because the concentration distribution becomes uniform for
the silts and clays. If surface sampling is used and
measurements are required to within 20% accuracy (for example)
then Figure 3.1 may be used to calculate limiting values of
$\bar{z}$ and hence limiting values of particle size D. Particles
smaller than this limiting value of size will then be
measured sufficiently accurately by the surface sampling
procedure.
Values of $\bar{z}$ for $qC_a/g_{ss} = 0.8$ read off from Figure 3.1.,
are listed in Table 3.2., together with the limiting values
of settling velocity w. The final line of Table 3.2. is
obtained from a standard plot of D versus w, shown in Figure
3.2. (see graf, 1972, p. 45 for extension of plot to larger
values of D).

Table 3.2.   Limiting sand sizes for use of a single sampling
              point to approximate the suspended sediment
              discharge concentration.

	Sampling Point					
	Mid-depth a/d = 0.5			Surface a/d = 0.90		
	1.5	2.5	3.5	1.5	2.5	3.5
Mean vel.V m/s	1.5	2.5	3.5	1.5	2.5	3.5
Shear vel.U_* m/s	0.15	0.25	0.35	0.15	0.25	0.35
$\bar{z}$ (from Fig. 3.1)	0.65	0.65	0.65	0.12	0.12	0.12
Settling vel. w mm/s	39	65	91	7.2	12	17
Limiting particle size D mm	0.35	0.50	0.70	0.10	0.14	0.17

.   Although the equations used to obtain Figure 3.1 and
hence Table 3.2. are approximations, and although von Karmen's
constant is less than 0.4 in sediment laden rivers, the trend
is clear. Mid-point sampling may be used to obtain reliable
values of g_{ss} under most conditions, and surface sampling may
be used as a satisfactory approximation provided the river

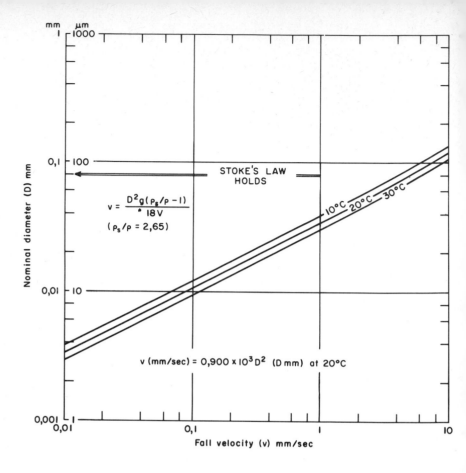

Fig 3.2. Settling velocity in water versus particle size

does not carry a heavy sand and gravel load. Because of the simplicity
of surface sampling procedures a very big reduction in field costs is
possible if surface sampling is used. This is recommended in the next
section as an operational method for certain kinds of rivers.

Where equipment which samples at many values of depth is used, it is
important (Leopold *et al.*, 1964, p. 186) to ensure that the equipment
works so as to give the product of concentration and velocity described
by equation 1. The suspended sediment discharge per unit width of stream
g_{ss} is not equal to the product $(\bar{c}\ \bar{u}\ d)$, depth average concentration
times depth average velocity times depth. g_{ss} is smaller than $(\bar{c}\ \bar{u}\ d)$
because the deviations of concentration and velocity from their mean
values are negatively correlated. Thus a suspended sediment sampler that
physically mixes samples from all values of depth must take amounts from
the various levels proportional to the velocities at the various levels
in order to give an aggregate suspended sediment concentration equal to
g_{ss}/q. Integrated suspended sediment samplers which take in equal amount
of water from the various levels will give final concentrations that are
too large. The magnitude of this error is significant for the sand size
only. For the silt and clay sizes the concentration versus depth
variation is sufficiently small that this effect is unimportant.

44

Fig 3.3. Manual withdrawal of large sample for size analysis

Equipment for suspended sediment measurement

Manual withdrawal of single samples This method involves an
observer dipping a wide mouthed bottle into the flow. The
bottle may be hand held for surface sampling, or may be
clamped at the end of a rod for deep immersion. Usually, the
observer samples from the river bank. The use of surface
sampling is shown in the previous section to cause significant
errors for the sand sizes, but small errors for the silt and
clay fraction. A site at which there is an established flow
gauging station with a recorder making continuous measurements
of the water discharge Q is selected. The suspended sediment
discharge G_{ss} is determined from

$$G_{ss} = QC \qquad\qquad (3.6.)$$

in which C is the concentration in the suspended sediment
sample.
    Use of manual sampling by resident observers is particularly
suitable in developing countries because it is usually
possible to employ observers who are patient and thorough at
reasonable cost. For small and medium sized catchments this
method is preferable to that using heavy samplers because
samples can be taken at very short notice, and short duration
floods can be monitored properly. In many tropical countries,
where floods on small and medium sized catchments arrive at
night from the previous afternoon's storm, the method is
advantageous. Occasional large (40 litre) samples are
gathered from reaches of very turbulent flow near the gauging
site (see Figure 3.3.) to check the ratio of sands to other
solids in the suspended sediment. After a period of one to
two weeks the water from the large sample is decanted and the
sample returned to the laboratory for size analysis.

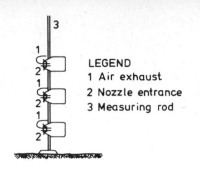

LEGEND
1 Air exhaust
2 Nozzle entrance
3 Measuring rod

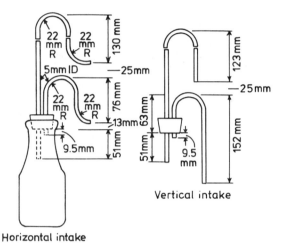

Vertical intake

Horizontal intake

Fig 3.4. Single stage surface sampler, US Federal Inter-Agency Sedimentation Project.

Sampling by this method has been successfully used in many parts of the world including Tanzania (Temple and Sunborg, 1972), and in Zimbabwe (Ward, 1980). Where the use of resident observers is prohibitively expensive, rivers may b sampled automatically on the rising stage using the simple bottle-syphon arrangement shown in Figure 3.4. Banks of several of these samplers, arranged at different heights ha been successfully used in the United States (US Interagency Committee on Water Resources, 1963).

Mechanical withdrawal of single samples Many designs of pumped sampler have been developed. Water samples are withdrawn from the river by an automatic sampler at known times, usually several times per day, and stored in bottles Occasional visits are made to the sampler to collect the filled bottles (to be taken for laboratory analysis) and to restart the sampler. The method requires a site close to a flow gauging station, so that Q is known and the suspended sediment discharge can be calculated from equation 3.6. Th method is suitable for rivers with small and medium sized catchments because it is able to measure rapid changes in flow by operating at high sampling frequencies. Usually th pump is attatched to the sampler at a point above the highe flood level. It should be remembered that the pump cannot

46

Fig 3.5. Sequential pump sampler, US Federal Inter-Agency Sedimentation Project, model US PS-69.

more than 9 m. (and some smaller distance at high altitudes) above the river surface otherwise it draws a vacuum. In the design shown in Figure 3.5., the pipeline from the inlet to the sampler is first primed using a small pump, pumping water in the reverse direction from a large container (seen to the right of the sampler) to the river. Then the main pump switches on and pumps for some time in the forward direction (while the flow discharges into the large container) so that the inlet pipe is filled with a fresh sample of river water. While the main pump is still operating, the flow is then switched so that it fills one of the plastic cups (seen in a tray underneath the sampler) and is finally switched back so that it fills the large container to the top. The microswitch to start the sampling cycle may be triggered at a variety of frequencies from once per 24 hours to 4 times per hour. In addition float switches are available which change the sampling frequency at large values of stage. The requirements for positioning a sampler of this kind at a river site are:

A) The sampler should be in a secure housing which will not be swept away during flood flows

B) The intake pipe should be sufficiently short that it is flushed thoroughly at the beginning of the sampling cycle

C) The sampler should be low enough to pump water at the lowest values of stage required and high enough to be above the maximum anticipated flood stage. Samplers of the type shown in Figure 3.5. are not suitable for use on rivers whose stage changes seasonally by more than 9 m, because this condition is not met.

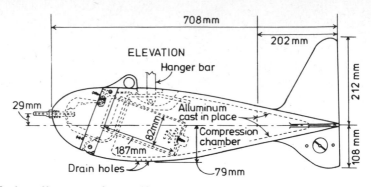

Fig 3.6. Heavy submersible sampler (model US P-61) for with drawal at multiple values of depth.

The entrance of the sampler intake should be carefully positioned if the pumped sampler is to measure the sand fraction of the suspended sediment correctly. The pipe entrance should face exactly upstream and the pipe diameter should be chosen so that the velocity in the pipe approxima the velocity of the river during significant flood events. At some sites it may be possible to make simultaneous measurements using pumped samplers and using multiple sampl on a few occasions, enabling future pumped sampler results be corrected with a calibration factor.

Multiple sampling across the section   Several samplers have been developed by the United States Federal Interagency Sedimentation Project for sampling suspended sediment from many points across a river section (Guy and Norman, 1970; US Interagency Committee on Water Resources, 1963). In all these designs the sample bottle is encased in a streamlined housing, with the intake nozzle extending forwards into the flow from the front of the housing. The samplers are lower from a bridge, cable, boat (large rivers) or by wading (sma rivers) so that many positions in the vertical and in the horizontal are sampled. Photographs and diagrams of the samplers have been published widely, see for example Guy an Norman (1970), Vanoni (1975), Graf (1972), Water Resources Council Sedimentation Committee, 1976. Figure 3.6. shows t US P-61 sampler (weight 48kg) designed for lowering by wincl from a bridge, cableway or boat. For fast flowing rivers t US P-50 sampler (weight 136 kg) is used. These samplers hav an electrically operated valve which is triggered from the surface when sampling is to commence. One design of sampler model US D-49, has no valve and samples continously while i is immersed. The suspended sediment transport is determined by the method of equally spaced verticals across the stream and an equal-transit-rate (ETR) at all verticals, or by the method of centroids-of-equal discharge increment (EDI) acros the stream (see Guy and Norman, 1970).

Provided the stream is not very deep (< 9m), vertically integrated samples are taken for both methods by lowering t instrument to the bed, opening the intake valve and allowing the sampler to fill as it ascends at a filling velocity equ to the local value of the stream velocity. The suspended sediment sample so taken has a concentration equal to the suspended sediment discharge concentration given by equatior

48

Sampling is carried out for many verticals and a number from
4 to 20 verticals is usually used (see Guy and Norman, 1970,
p. 41). Summation horizontally across the section (EDI method)
is carried out by multiplying the concentrations (measured by
laboratory analysis) by the fractional discharge and adding.
Thus, the EDI method requires water discharge measurement
simultaneous with the sediment sampling. The ETR method takes
a sample proportional to the amount of flow at each of several
equally spaced verticals in the cross section. These samples
are combined, and a gross sample proportional to the total
stream flow is taken, obviating the need for simultaneous
water discharge measurements. The ETR method has the
additional advantage that samples from the whole section are
integrated in one gross sample, reducing laboratory analysis
time and costs. Care must be taken with both the EDI and the
ETR methods that the sampler is lowered at a rate significantly
lower than the velocity of flow. As an approximate guide, the
transit rate of the sampler should be less than 40% of the mean
velocity. Problems to be avoided in the use of the samplers
are: air compression in the sampling bottle, nozzle size being
too small, and major deviations of the angle of the suspending
line from the vertical at large velocities (Guy and Norman,
1970). The instruments are suitable for measurements in large
rivers where changes of stage do not occur rapidly. Their use
is desirable in large rivers which carry large proportions of
sand, because measurements are made at all depths, including
close to the bed. They are unsuitable for use on small rivers
with rapid change of stage, and on rivers in arid and semiarid
areas because it is usually not possible for the measuring team
to catch the (vary rare) significant flood events.
Direct reading from in situ instruments Suspended sediment
concentrations may be measured by permanently installed
instruments in the river which rely on the principle of light
or gamma ray attenuation. Instruments which monitor the
attenuation of a light beam with a photo-electric cell are of
the direct immersion type, or are mounted on the bank and rely
on a pump to circulate a sample of the river water through the
photoelectic cell. The direct immersion cell method (developed
by United Kingdom Water Pollution Research Laboratory) has been
in operational use for several years in the United Kingdom
(Fleming, 1969; Thorn, 1975). Instruments using both single
path and dual path light beams (see Figure 3.7.) are in use.
With the dual path method, the ratio between the light
intensities reaching the two photo-electric cells (see also
equation 3.10) is:

$$\frac{I_{p2}}{I_{p1}} = \exp\left(KC(l_1 - l_2)\right)$$

in which C is the concentration of suspended sediment, l_2 and
l_1 are the path lengths and K is a calibration factor which
depends upon particle shape, particle light absorption
properties, particle size distribution and particle specific
gravity. The dual path, unequal gap method is a useful
technique of removing errors due to clouding of lenses and
errors due to ambient diffused light. Another design (the
Process Control Photometer, manufactured by Euro Technology
Inc.) includes two lamps and two photocells arranged in a

49

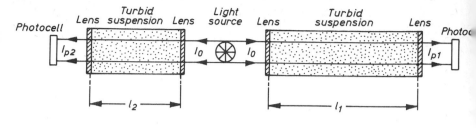

DUAL-PATH COMPENSATION

THEORY OF OPTICAL SUSPENDED SOLIDS MONITORS

Fig 3.7. Determination of suspended sediment concentration light attenuation (turbidity) measurements.

special geometric configuration. This allows the effects of lamp and cell sensitivity changes, as well as contamination of windows, to be eliminated. The sensing head is clamped to a bridge pier or to a flow gauging structure near the centre of flow at approximately mid-depth and is connected by a length of good quality marine cable to the monitor/recorder system on the bank, above the highest flood level. The instrument shown in Figure 3.8. is battery operated, has a clock operated switching system (not shown) which activates the monitor for a brief period once every hour, and carries sensing head (see Figure 3.8.) suitable for concentrations of up to 2 000 mg/l. Sensors for both larger (using a single light path) and smaller values of concentration have been developed. The head is removed for cleaning at intervals. The cleaning interval depends upon the operating conditions a period of about 1 week is satisfactory in rivers which do not carry very heavy algal and weed concentrations.

The instrument's sensitivity and zero position is checked before and after cleaning with a standard solution. Formazin (a mixture of hexamine and hydrazinium sulphate) may be used The particle size is the most important of the various factor that affect the calibration of factor K, and large differenc in K arise for silt sized particles compared with clay sizes

For this reason it is necessary to establish individual calibration curves for the suspended sediment monitor which apply to rivers draining each of the various types of catchm being studied. The calibration curve for the instrument, obtained using formazin solution, is given by Kiff (1977). Also given are factors for the conversion of the formazin calibration to the true calibration, obtained for suspended sediments from several rivers and estuaries. The method is sensitive to the positioning of the measuring head below the

Fig 3.8. Suspended solids monitor with dual path sensing head, (Partech Electronics Ltd., St. Austell, England).

51

water surface for rivers which carry a large fraction of sand sizes. For rivers which carry mainly clay and silts it is possible to operate the instrument near the surface or near the bed (see Figure 3.1.). The measurements must be made at a flow gauging station with continuous recordings so that the suspended sediment load can be determined by equation 3.6. A comparison between the performance of this method with the probe in the lower half of the flow and the results of manual sampling near the surface followed by laboratory analysis (Ward, 1978), on a river where the sand fraction was a small percentage of the total suspended sediment, showed good agreement between the two methods.

Light attenuation suspended sediment meters of the cell type, through which a sample of river water is pumped, are in operational use at a few locations in North America.

Experimental work has been undertaken with various design of suspended sediment gauges that rely on the attenuation of X-rays from a source of Cadmium-109 (McHenry *et al.*, 1967; Murphree *et al.*, 1968). Although the method showed promise it has not been widely adopted for operational use. The penetration of X-rays through water is such that unless the suspended sediment concentration is large, the reduction in signal is small, resulting in poor sensitivity at low concentrations.

Bed load measurement

Bed load is the part of the sediment load in which the particles move by rolling and sliding along the bed. In this section, methods are described which rely on field measurement alone, or on both field measurement and calculation to determine the bed load discharge.

Channel and bed material measurements followed by calculation
Einstein's calculation (Einstein, 1950) for bed material load enables the bed material load in an alluvial reach of a sand bed river to be predicted. Einstein (1950) describes an alluvial reach as a reach in which the river is neither depositing nor scouring. In the test reach selected, the shear stress of the boundary of the moving water works only against the sandy bed of the river. Rock outcrops do not occur, or are insignificant, no significant tributaries enter the river, and no islands occur.

Measurements of the average cross-sectional area of the channel and mean gradient of the test reach are made by carrying out a survey at times of low flow. Bed material is collected from the sand bed, also at times of low flow, care being taken to take representative samples. Guidelines for sampling dry river beds are given by Einstein (1964). The bed material samples are sized in the laboratory by sieving. The test reach should be close to a flow gauging station, so that the frequency of various sized flood events is known. The first part of the procedure is a calculation of the hydraulic characteristics (stage, velocity, discharge) of the flow in the reach, for assumed values of the hydraulic radius Likely values of the hydraulic radius are found by trial such that the maximum value gives a stage equal to the bankfull value. Usually about six to eight values of hydraulic radius are assumed and values of stage, velocity and discharge are

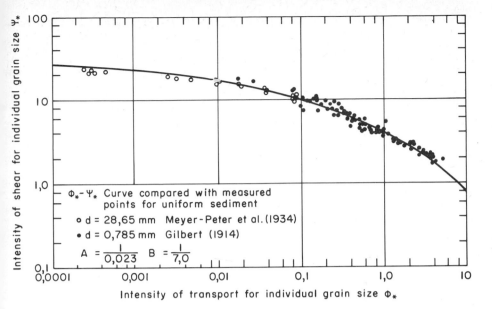

Fig 3.9. Intensity of transport versus intensity of shear for bed material load (after Einstein, 1950).

calculated for each (see Einstein, 1950 and Graf, 1972, p.224).
     Using the size analysis of the bed material, about six ranges of sand sizes are selected to cover the whole distribution (the limits of the sediment range usually being taken as those sizes for which 98% and 2% of the bed material is respectively smaller and larger). The bed material load for each of the grain sizes for each value of river flow is then calculated using a graph (Figure 3.9.) which relates ϕ_*, the dimensionless intensity of transport for an individual grain size, to ψ_*, the dimensionless intensity of shear on an individual grain size. These are given by the equations:

$$\psi_* = \xi Y \left(\beta^2/\beta_x^2\right) \frac{p_s - p_f}{p_f} \frac{D}{R_b' S_e} \qquad (3.7.)$$

and $$\phi_* = \frac{i_B}{i_b} \frac{q_B}{p_s g} \left(\frac{p_f}{p_s - p_f}\right)^{\frac{1}{2}} \left(\frac{1}{gD^3}\right)^{\frac{1}{2}} \qquad (3.8.)$$

in which ξ = hiding factor of grains in a mixture (given by Einstein, 1950)

Y = pressure correction in transition rough-smooth given by Einstein, 1950)

(β/β_x) = logarithmic function (given by Einstein, 1950)

p_s & p_f = densities of sediment and fluid respectively

D = grain size of fraction being calculated

R_b = hydraulic radius associated with shear on sand grains

S_e = slope of energy grade line

i_B = fraction of bed load in given grain size

i_b = fraction of bed material in given grain size

q_B = bed load rate in weight per unit of time and width

g = gravitational acceleration

The relationship $\phi* $ = function $(\psi*)$ used by Einstein was initially based on experiments for grains which varied over wide range of sizes (Einstein, 1942) and subsequently based Einstein's bed load theory (Einstein, 1950), which was show to be in good agreement with the original experimental resu The total sediment transport rate for the particle size fraction is determined by summing the bed material load mov in suspension and the bed material load moving as bed load:

$$i_t q_t = i_B q_B (PI_1 + I_2) + i_B q_B \qquad (3.9.)$$

in which i_t = fraction of total load in a given grain size

q_t = total bed material load per unit width per un time

P = dimensionless parameter of total transport (s Einstein, 1950)

I_1, I_2 = integrals given graphically by Einstein (1950

Integrals I_1 and I_2 are functions of the settling velocity the particle and are obtained knowing the concentration of sediment at the bed by a similar procedure (Einstein, 1950) the integration used previously to obtain equation 3.5. Finally, the total transport rate for all grain sizes is fou by adding the transport rates for the individual size components of the bed material.

Einstein's method has the advantage of being applicable rivers which change stage quickly. There is no requirement that measurements be made during the flood for which information is required. Measurements are made during times of low flow, and the predicted sediment transport is compute later. The calculation can be made by one person in less than a week and in much shorter time if parts of the calculation are run by desk-top computer. One of the disadvantages is that errors in the hydraulic calculation ma cause much larger errors in the sediment transport of the calculation. A comparison of Einstein's method with a set of field measurements for the Niobrara and Colorado rivers is given by Vanoni (1975, p. 221) and shows that the Einstein method underestimates the bed material load by a factor of between 2 and 4. Graf (1972, p. 219) points out that the bed material load calculated by this method is subject to errors in the various coefficients used, not all of which are known accurately. Graf concludes that these uncertainties are responsible for errors of as much as 100% in the final calculation.

The modified Einstein method was developed by Colby *et a* (1955, 1961) for measuring the total load (bed material plus wash load) in rivers. Its use is confined to cases where the suspended sediment concentration during the flood is measured. The stream discharge, mean velocity, cross-sectional shape, sediment-size distribution of the suspended sediment, and sediment-size distribution of the bed sediment are required. The fact that the velocities used in the calculation are measured values is a disadvantage over the basic Einstein method, since the large floods required for engineering predictions are frequently impossible to gauge. However, if data on velocity and on the other required available, then the application of the modified method is advantageous, because the use of measured, rather than predicted, velocities is a more reliable approach. Einstein (1964) presents a much simpler version of the integral required to determine the total load than did the authors in their original papers, leading to exactly the same final result. The modified Einstein method is accepted by several authorities (Vanoni, 1975; Nordin, 1977) as being the most accurate procedure available today for measuring the total discharge of sand sized material.

Bed load samplers   A suitably designed trap is lowered to the bed during times of significant flood flows and allowed to collect a sample of the bed load. Although this objective sounds straightforward, most designs of bed load sampler have not been satisfactory. Three major problems prevent good performance: (a) disturbance of the flow and hence the pattern of bed load movement by the sampler, (b) the fluctuating nature of the bed load transport, and (c) improper operation, resulting in the sampler digging into the bed layer. The first problem, (a), causes many designs of sample to take in less bed load than the true bed load rate. Efficiencies of several early samplers tested by Novak (see Graf, 1972, p.360) are in the range 40 to 60%.

One of the most successful samplers is the Arnhem sampler developed in Holland at the Delft Hydraulics Laboratory (Figure 3.10.). Its hydrodynamic design causes the velocity of entry into the intake to be equal to the external velocity. It is robust and heavy (32 kg), and must be lowered to the streambed by winch. A leaf spring attached to the frame presses the sampler against the bed. The mouth has an opening 85mm x 50mm. Flow exits through a wire mesh trap, whose mesh size is normally 0.3mm. At the end of the sampling interval, the sampler is raised to the surface, the wire mesh trap removed, and the bed load sample is washed into a container for laboratory analysis. Repeated samplings at the same place and at several positions across the section are required to attempt to smooth out the fluctuations in bed load transport. Hubbell (1964) reports temporal fluctuations in bed load collected in a sampler of 11 ml to 112 ml per minute on the Middle Loup River, Nebraska, USA, and Bagnold (1977) reports large fluctuations in transport on the East Fork River, Wyoming, USA.

The Arnhem sampler can be used only in small and medium flow velocities (up to 1.5 m/s). At high velocities, the sampler cannot be lowered to the bed. Recently a bed load sampler developed by the United States Geological Survey, the Helley-Smith sampler, has been successfully used in measurements on the Snake and Clearwater rivers in the United States (Emmett, 1976; Bagnold, 1977). The sampler was

55

Fig 3.10. Arnhem sampler for bed load measurement, double frame collapses slowly as rope tension is released, causing sampler inlet to be gradually lowered onto the bed, (Van Essen Co., Delft, The Netherlands).

calibrated using a specially designed bed load weighting system installed in the East Fork River, Wyoming (see chapter 5). The Helley-Smith sampler was found to provide satiafactory measurements as long as systematic traverses of the river bed were repeated several times and the results averaged. In this way the effects of streams of solids whic wander at random across the bed were averaged out.

Several types of bed material samplers have been introdu by the United States Government Federal Inter-Agency Sedimentation Project (Guy and Norman, 1970). These are all designed to be used for taking grab samples of bed material. Samples of approximately 175 ml, penetrating about 40mm into the bed, are taken. These samplers do not measure the bedload discharges. Similar samplers are in use in the Netherlands and all over the world (Van Veen's Grab, manufac by Van Essen Co., Delft, The Netherlands).

Slot traps   Traps set in the bed of small rivers have also been used to determine the bed load transport. In 1941, the United States Department of Agriculture, Soil Conservation Service, made measure-measurements with slot traps, from which sediment was pumped to a tank for volumetric measureme (Vanoni, 1975, p. 344). Recently (1973), the United States Geological Survey constructed a bed load measuring system on the East Fork River, Wyoming (Leopold and Emmett, 1976). Hydraulically- operated horizontal gates opening to 150mm width and flush with the bed surface, were installed across the 15mm width of channel bed. Bed load entering the slot w carried by conveyor belt, raised by bucket lift and weighed batches, before being returned to the river. The gates were operated so that the bed load was trapped either in individu sections or across the whole width simultaneously.

This method is not in routine use because it is obviously expensive to operate. However, the method is important as it provides a way of calibrating bed load samplers and of studying the temporal and spatial fluctuations of bed load under prototype conditions.

The Institute of Hydrology (Wallingford), UK, has been conducting research in the Plynlimon catchments, Wales, on the movements of cobble size bedload using large concrete pits in the bed of the river, emptied using a mechanical shovel. Fraction collectors, e.g. slot divisors, have also been used on small rivers in various parts of the world, usually for soil erosion research. A known small percentage of the river discharge is diverted during the flood and stored, and the sediment is allowed to settle. Volumetric or gravimetric measurements enable the total amount of sediment carried during the flood to be determined.

Tracers   Tracing of bed load movement may be carried out using naturally occurring contrasts in bed material characteristics or using artificial labelling. Artifical labelling has been undertaken using paint (for large size bed material), fluorescent dyes (Teleki, 1963) and radioactive elements (Crickmore, 1967; Crickmore and Lean, 1962). An experiment in the North Loup River, Nebraska, USA (Hubbell and Sayre, 1964) using sand labelled with Iridium 192 gave a value for bed load transport in good agreement with that computed using the modified Einstein method. A sled-mounted scintillation counter was used for detecting the tracer. Hubbell and Sayre used the mean velocity of movement of the centroid of the tracer cloud, and various methods of determining the depth of the zone of particle movement, to find the bed load discharge. A restriction of the method is that the measurements must be made over an extended period of time (of the order of days). The chance that the flow conditions remain approximately constant during such a long period is small. In an application in the Firth of Forth estuary, Scotland, Smith and Parsons (1967) used Scandium 46 incorporated in ground glass which was crushed until its size distribution matched that of the estuary silts, to demonstrate that dredging practices in use in that estuary were inefficien and that different dumping grounds for dredge material were required. Present day restrictions about pollution of rivers and estuaries mitigate against the use of radioisotopes for these measurements even though isotopes of very short half-life are available.

LABORATORY METHODS

Introduction

The primary objective of the analysis of river sediments in the laboratory is to determine the concentration of the suspended sediment in the samples. In addition, measurement of the size distribution of the suspended sediment is important. Additional measurements, such as particle shape, specific gravity of particles and heavy mineral proportion are sometimes required and methods of measuring these quantities may be found in standard textbooks on sedimentary

petrology (e.g. Engelhardt, Fuchauer and Muller, 1967).
Measurement of dissolved solids is described in Chapter 6.
    Fluvial sediments range over a very wide group of
particle sizes, as shown in Table 3.1. The properties of
these particles are functions of their sizes. Table 3.3.
(adapted from Guy, 1969) describes the physical properties
of four categories of fluvial sediment. The sizes of other
physical quantitites, such as the diameter of bacteria and
the wavelength of visible light are included for reference.

Suspended sediment concentration

    The two usual methods of determining suspended sediment
concentrations, filtration and evaporation, are standard
procedures (Guy, 1969). A third method, using the turbidity
of the solution measured with a light transmissibility cell
is not in general laboratory use but merits more attention.
    The accuracy of most closed pan laboratory balances
($\pm$ 0.25 mg) puts a lower limit on the minimum suspended
sediment concentration that can be measured. Assuming the
minimum weight of dried sediment for reasonable accuracy is
5 mg and assuming the sample volume collected from the river
is 500 ml, then the lower limit of concentration measurable
in routine analysis is 10 mg/l. This is a small value, most
fluvial sediment concentrations being in the range 50 to
5000 mg/l, and thus the weighing accuracy is not usually
a constraint on the measurement.
    The filtration method is satisfactory for samples that
are predominantly clay, only if the concentration is small.
In practice it is found that the concentration must be less
than 200 mg/l, otherwise the filter becomes blocked. With
samples that are mostly sand (for example a sample of bed
material) filtration may be used up to large concentrations.
    The evaporation method is complicated by the fact that
the dissolved solids are weighed together with the suspended
solids after evaporation. A correction may or may not be
needed in determining the weight of suspended solids.
Evaporation   This method involves allowing the sediment to
settle to the bottom of the sampling bottle, carefully
decanting the supernatant liquid, washing the sediment into
an evaporating dish, and drying the sediment in an oven.
The settling process takes several days with samples that
have large clay concentrations. For this reason, special
procedures for speeding up the settling process, such as
centrifuging or flocculating agents, may be used (Guy, 1969,
p. 11).
    A small quantity (about 25 ml) of supernatant liquid is
left with the sample to avoid losses. For a measuring error
of less than 5%, a dissolved solids correction should be made
when the dissolved solids concentration is equal to or
greater than the suspended solids concentration in the
original 500 ml sample. In routine analysis this correction
will frequently not be necessary because the suspended solid
concentration is usually greater than the dissolved solids
concentration. The evaporating dishes are dried at a
temperature low enough to prevent spattering of sediment from
the dishes by boiling action ($80^{\circ}C$ is satisfactory). After
loss of all visible water, the dishes are heated at about

Table 3.3.   Physical properties of small particles

	Coarse Suspensions	Colloidal Suspensions	Colloidal Solutions	Molecular Solutions
ze Range, , microns	2 - 1000	0.1 - 2	0.001 - 2	0.001 - 0.0001
pearance in ter	Very cloudy	Turbid	Clear	Clear
rticles Observed th:	Naked eye & microscope	Microscope & electron microscope	Electron microscope	Cannot be Observed
te of settling	Quickly or overnight	Slowly or not at all	Do not settle	Do not settle
rticles separated om water by:	Filter paper	Membrane filtration	Membrane (nuclepore) filtration	Not by filtration
termination of rticle sizes	Sieves, microscope, gravity or centrifugal settling	Electron microscope, centrifugal settling, light settling	Ultra-centrifuge	Ultra-centrifuge
rm after aporation	Loose powders	Powders and gels	Gels	Crystals

imits of size measurements of microscopes and filtering materials:

Microscope,	down to 2 μm
iltration, paper filters	down to 2 μm
" membrane filters	down to 0.2 μm
" nuclepore filters	down to 0.03 μm
avelength of red light	0.7 μm
" of blue light	0.45 μm
acteria	0.5 to 1.2 μm

110°C for some time to ensure complete drying. The final
value for concentration in mg/l is found by dividing the
weight of suspended sediment by the original volume of sampl
Provided the concentrations are less than 16 000 mg/l the
numerical value of concentration in parts per million by
weight closely approximates the value in mg/l.
Filtration   The samples are filtered under vacuum through a
Gooch crucible or Buchner funnel fitted with a glass-fibre
filter disc. These filter discs show no weight loss during
filtration. The crucibles and filters are placed in the
drying oven and dried using the same procedure described for
the evaporation method. The containers and filters are cool
in a dessicator at room temperature prior to weighing

    For samples containing fine sediment (clays) it may be
necessary to use an asbestos mat on top of the glass-fibre
filter disc to prevent blockage. The mat is prepared in the
standard way used in chemical analysis, by pouring an asbest
slurry on top of the disc while vacuum is applied. During t
filtration process the asbestos mat retains much of the
sediment that would ordinarily clog the glass-fibre disc.

Fig 3.11.   Instrument for turbidity measurements in the
laboratory, showing three sizes of measuring cell.

The disadvantage of this method is that the asbestos mats ha
to be individually prepared by filtration of the asbestos
slurry, rinsing, oven drying and cooling in a desiccator.
These steps make the time required for a single analysis
large. Membrane filters may also be used when low suspended
sediment concentrations are involved and when it is necessar
to recover fine clay-sized particles (cf. Table 3.3.).

<u>Light transmissibility</u> A small volume of sample is poured into a test tube and the turbidity is measured in a laboratory colorimeter. The colorimeter is an inexpensive laboratory instrument (see Figure 3.11.) consisting of a light source, a measuring tube, and a photoelectric cell. The principle of operation is the same that for all turbidity measurements. The ratio between the light intensity I transmitted through the sample and the light intensity I_o transmitted through distilled water is given by:

$$I/I_o = \exp(KC^n) \qquad\qquad (3.10.)$$

where C is the concentration of the suspension

n is a constant

K is a parameter which depends upon instrument design and particle size

n is found to be equal to 1.0 for calibrations at small values of concentration. For calibrations extending from small values up to large values of concentration (over a span of about 2 orders of magnitude) a value of n smaller than 1.0 is necessary.

The method works well for samples with large proportions of clay. The turbidity parameter K, a measure of the sensitivity of the method, is strongly dependent on the particle size. K is about an order of magnitude larger for fine sands than it is for fine silts, because the ratio of particle surface area to particle volume becomes larger as particle size decreases. K becomes approximately constant in the clay sizes. Despite this variability of K, it is found in practise that a single calibration curve (implying the same value of K) can be used for all suspended sediment samples from a given river catchment within reasonable error bounds ($\pm$ 10%). In some cases, groups of rivers all draining catchments with similar geology will have the same suspended sediment calibration, K. Although this method is not in general use, possibly because the concentration results are not precise, the method warrants more attention. Colorimeter analysis is fast, uses little laboratory equipment, requires very small volume samples (only about 2.5 - 15ml), and is self checking, because the operator can see immediately whether the reading on the colorimeter is about right from his visual impression of the sample. Selection of a suitable sized measuring cell means that almost all the concentrations of interest can be measured without preliminary dilution. Calibrations are carried out by measuring samples of known concentration (found by evaporation or filtration). The method gives results to an accuracy compatible with the accuracy to which the water discharges are measured. Samples containing the large percentages of sand cannot be measured because the sand falls to the bottom of the sample tube before a reading can be made.

Size distribution

Sediment-size distributions are commonly determined by sieving for the sand sizes and settling in a water column for the silts and course clays. These methods, and other less common methods, are discussed in this section. There is no single method that satisfactorily covers the whole range of

61

river sediment sizes. The sizes obtained are related to the
method of measurement and it is difficult to relate precise
the results of different measuring techniques. For example
particles of sieve diameter less than 75 μm may have fall
velocities corresponding to a stokes diameter of up to 100
(Kiff, 1973). Additional problems of measurement by settli
in water columns may arise for clay particles because of
flocculation. In hydraulic engineering problems the fall
velocity is normally the parameter required in calculation.
Thus, methods employed settling in a water column are gener
used where possible for size analysis. The standard fall
diameter of a particle is the diameter of a sphere that has
a specific gravity of 2.65 and has the same fall velocity a
the particle.
Fall velocity Particle sizes in the silt and coarse clay
category are commonly determined by pipette analysis or
bottom withdrawal tube using settling in a long column of
water (see Figure 3.12.). There is no clear overall advant
in the use of one of these methods over the other. The US
Geological Survey (Guy, 1969) recommends pipette analysis f
concentrations of 2000 to 5000 mg/1 whereas bottom withdraw
tubes are used in the range of 1000 to 3500 mg/1. Kiff (19
uses bottom withdrawal tubes for the range 2000 to 6000 mg/
All methods use a suspension which is initially well mixed.

In the pipette analysis method, 25 ml samples are
withdrawn from a height, L., either 50, 100 or 150mm below
surface of the suspension at predetermined values of time,
after the start of the test (see Guy, 1969, p. 24). During
time t all particles larger than a given size have fallen m
than distance L, and particles smaller than this size remai
at height L. These particles that remain are measured in t
sample drawn into the pipette. A table of values for weigh
of particle remaining in suspension for various values of
settling velocity (L/t) is drawn up, and these values of
settling velocity are related to the standard fall diameter
of the particles using Figure 3.2. Thus, a cumulative curv
(percent less than or percent greater than) of frequency
versus particle diameter is prepared.

With bottom withdrawal analysis, particles which have
fallen the length of the water column are washed into the
beaker at times t_1, t_2, and t_3 after the start of the test
(Figure 3.13.). Samples of volume about 50ml are collected
and are analysed for weight of suspended sediment. A graph
percent remaining in suspension versus time of fall, t, (kn
the Oden curve, Figure 3.12.) is prepared and values of
frequency of occurence of the various sizes of sediment D
(corresponding to the various settling velocities) are foun
from the gradient of the Oden curve. At some value of time
(e.g. t_1) particles of settling velocity v_1 have not quite
ceased to settle on the bottom of the withdrawal tube. The
rate of accumulation at t_1, is the same rate that would hav
arisen if particles of size characterised by velocity v_1, a
smaller were accumulating. Larger particles have no effect
because they have all settled out already. After sampling
time t_1, the distance of fall is L_1, and thus $v_1 = L_1/t_1$. B
drawing a tangent back until it reaches the ordinate line it
is possible to determine the mass M_1 (expressed as a percent
of all sizes of particles which contribute to the gradient

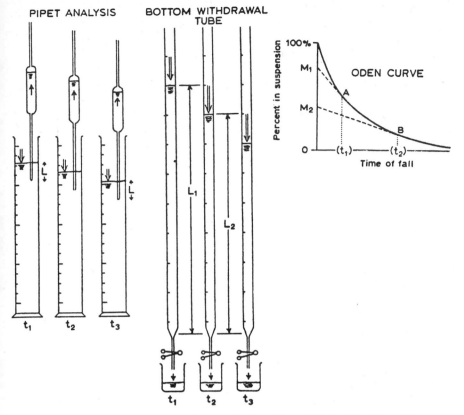

PIPET ANALYSIS   BOTTOM WITHDRAWAL TUBE

ODEN CURVE

Percent in suspension

100%

M_1

M_2

A

B

(t_1) (t_2)

Time of fall

t_1 t_2 t_3

L_1

L_2

t_1 t_2 t_3

Fig 3.12. Measurement of particle size distribution by settling in a water column.

at point A. Drawing a tangent from some other point B determines the mass M_2 of particles which have a settling velocity less than or equal to (L_2/t_2). This is the principle behind the use of the Oden curve.

Samples are first wet sieved with a fine (0.062mm) sieve to remove the sand and coarse matter. Organic matter that passes the sieve is removed from the sample using oxidation by hydrogen peroxide (see Kiff, 1973, p. 261). Splitting or dilution of the wet silt-clay samples is carried out to bring the concentration of the suspension into the correct range. A dispersing agent (sodium hexametaphosphate) is added, the sample is mixed and is then transferred to either a bottom withdrawal tube or a measuring cylinder for analysis. Details of exact quantities to be used in sample preparation, settling times and specifications for the bottom withdrawal tube are given in many sources, for example Kiff (1973), and Guy (1969).

Pipette analysis and bottom withdrawal tube measurements are suitable for particles in the silt and coarse clay ranges only. Both methods become unreliable for particles smaller than 2μm because convection currents in the settling medium become significant, and because of Brownian motion. Sands may be analysed either by sieving, or by use of settling in a visual accumulation tube (Guy, 1969, p. 30).

Several other methods of determining particle size

Fig 3.13. Bottom withdrawal tubes for particle size analysis
distributions by settling are available but space prohibits
full discription. The hydrometer method is used in soils
analysis, the Mettler sedimentation cylinder (direct weighing
is a variant of the bottom withdrawal method and the disc
centrifuge (Kiff, 1973) is suitable for measuring very small
particles (down to 0.1μm).

Sieving For the portion of the dry sample greater than
0.075mm diameter, a mechanical vibrating nest of sieves is
used to separate the various size components. An air-jet
sieve may be used for the small sizes, usually from 0.075mm
down to 0.038mm diameter. This provides some overlap with
results obtained by fall velocity determinations.

It has been shown (see data in Kiff, 1973) that there is
no attainable end point in sieving. A continual increase w
time in material passing the sieve is noted. The cause of t
increase is partially that of breakdown of particles and ch
because the smaller particles which remain on the sieve cha
and eventually find, the largest apertures in the sieve mes
Thus, a standard sieving time has to be used. The load on
sieves has a marked effect on sieving results. Using a mix
of very small glass beads, Kiff (1973), shows that the use
a small sample (10g) instead of a large sample (200g) causes
the final results of the analysis to be biased towards the
larger sizes. Small diameter (100mm) sieves are available
the range 0.075 to 0.005mm but these are not usually used i
routine sediment analysis.

Another method of sieving which deserves more attention
is the use of membrane filtration (see Figure 3.14.). These
filters have been developed for medical research, for remov

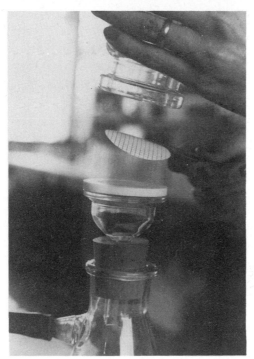

Fig 3.14. Membrane filtration equipment for particle size analysis of very small particles.

of bacteria of various sizes from water. Recently a new kind of membrane filter, a nuclepore membrane, has been introduced with holes of carefully controlled size penetrating the membrane. A variety of pore sizes in the range 0.03 to 8μm is available. Providing care is taken to ensure that these membranes are not overloaded and blocked, they can be used for separating fine silts and clays by filtration. As long as the layer on the membrane is no larger than a few particles thick, blocking does not occur. Filtration through a set of these filters of different sizes can be carried out, analogous to sieving through a bank of sieves. Although this method is not in use it is possible that it would provide a feasible method of sizing clays in the range 2 to 0.2μm.

Other methods   Electrolytic measurement of particle size has been developed as a routine method in the medical field. Equipment (i.e. the Coulter Counter) is commercially available, (see Figure 3.15.) and may be used for sediment sizing. The instrument involves circuitry that registers the changes in current between two electrodes as particles from the suspension are drawn through a small aperture. The sizes of the fluctuations in current are proportional to the volumes of the particles. A very fast electronic counter enables all particles in a given small volume of sample (e.g. 0.5ml or 2ml) to be registered. Upper and lower electrical thresholds are applied and are varied so that the number of particles in a predetermined size range may be found. The size distribution of the particles in the suspension may thus be found. The sample is mixed with an electrolyte (isoton solution) prior to

65

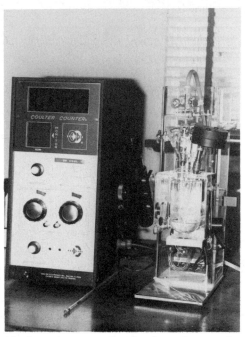

Fig 3.15. Particle size analysis by electrolytic current/ electronic counter measurements (Coulter Electronics Inc. Hialeah, Florida, U.S.A.).

analysis. Further work is needed to see whether the method effective for clay sized particles. Possible coagulation of the fine fraction of the suspended particles by the electrol may render this method unsuitable. Suspended material in se water may be analysed directly, using the sea water itself a the electrolyte (Sheldon and Parsons, 1967). The range of particle sizes measurable extends from about 150µm down to 0.6µm.

Recently light scattering techniques have been developed for industrial applications where mixtures of particle sizes are to be measured. These methods do not determine particle size distributions, but measure only mean properties of the mixture of particles. Cheesman (1978) has developed a metho of measuring the average particle size of a mixture by turbidity measurements of a suspension at a variety of light wavelengths. Wertheimer and Wilcock (1976) use specially designed Fraunhofer masks to determine the area mean radius, area standard deviations and the volume mean radius of a mixture of particles. Wertheimers instrument is used industrially to moniter particles from grinding processes.

ACKNOWLEDGEMENT

The author gratefully acknowledges the encouragement and assistance provided by the department of Civil Engieering, University of Zimbabwe, in preparing this manuscript whilst was on their faculty.

REFERENCES

Bagnold, R.A., 1977, Bed load transport by natural rivers, *Water Resources Research*, 13, 303-312.

Brooks, N.H., 1965, Calculation of suspended load discharge from velocity, and concentration parameters, *Proceedings of the Federal Inter-Agency Sedimentation Conference, 1963*, Miscellaneous Publication 970, Agricultural Research Service, US Department of Agriculture, Washington, D.C., 229-237.

Carter, R.W., and Anderson, I.E., 1963, Accuracy of current meter measurements, *Proceedings, American Society of Civil Engineers, Journal of the Hydraulics Division*, 89, 105-115.

Cheesman, G.C.N., 1978, A rapid method for the measurement of the average particle size of non-uniform lattices and dispersions, in *Particle size analysis, Proceedings of a conference organised by the Analytical Division of the Chemical Society, September 1977*.

Church, M., 1972, Baffin Island sandurs: a study of Arctic fluvial processes, *Geological Survey of Canada Bulletin* No. 216.

Colby, B.R., and Hembree, C.H., 1955, Computations of total sediment discharge, Niobrara River near Cody, Nebraska, *US Geological Survey Water Supply Paper* 1357.

Colby, B.R., and Hubbell, D.W., 1961, Simplified method for computing total sediment discharge with modified Einstein procedure, *US Geological Survey Water Supply Paper* 1593.

Crickmore, M.J., 1967, Measurement of sand transport in rivers with special reference to tracer methods, *Sedimentology*, 8, 175-228.

Crickmore, M.J., and Lean, G.H., 1962, The measurement of sand transport by means of radioactive tracers, *Proceedings of the Royal Society*, Part A, 266, 402-421.

Einstein, H.A., 1942, Formulas for the transportation of bed load, *Transactions, American Society of Civil Engineers*, 107, 561-597.

Einstein, H.A., 1964, River sedimentation, in *Handbook of Applied Hydrology*, ed Chow, Ven, Te, (McGraw-Hill, New York), 17-35-17-67.

Einstein, H.A., 1969, Sedimentation problems in Turkey, *Report to the General Directorate of the State Hydraulics Works, Ankara, Turkey*, October 1969 (unpublished).

Einstein, H.A., 1950, The bed-load function for sediment
transportation in open channel flows, *US Department
of Agriculture, Soil Conservation Service, Technical
Bulletin* No. 1026, reproduced in *Sedimentation,* ed
Shen, H.W., (1972) (H.W. Shen, Fort Collins, Colorado
USA)

Emmett, W.W., 1976, Bed-load transport in two large gravel-
bed rivers, Idaho and Washington, *Proceedings of the
Third Federal Inter-Agency Sedimentation Conference,*
1100-1115.

Engelhardt, W. v., Fuchtbauer, H., and Muller, G., 1967,
*Sedimentary petrology, Part I, Methods in sedimentary
petrology,* (Hafner Publishing Co., New York).

Fleming, G., 1969, Suspended solids monitoring: a comparison
between three instruments, *Water and Water Engineering*
337-832.

Gilbert, G.K., 1914, The transportation of debris by running
water, *US Geological Survey Professional Paper* 86.

Graf, W.H., 1972, *Hydraulics of sediment transport,* (McGraw-
Hill, New York).

Guy, H.P., 1969, Laboratory theory and methods for sediment
analysis, *Techniques of Water Resources Investigations
of the United States Geological Survey,* Book 5,
Chapter C1.

Guy, H.P., and Norman, V.W., 1970, Field methods for
measurement of fluvial sediment, *Techniques of Water
Resources Investigations of the United States
Geological Survey,* Book 3, Chapter C2.

Hays, F. ch., 1978, Guidance for hydrographic and hydrometric
surveys, *Publication No. 200, Delft Hydraulics
Laboratory,* Delft, The Netherlands.

Hubbell, D.W., 1977, Apparatus and techniques for measuring
bed load, *US Geological Survey Water Supply Paper*
1748.

Hubbell, D.W., and Sayre, W.W., 1964, Sand transport studies
with radioactive tracers, *Proceedings, American Society
of Civil Engineers, Journal of the Hydraulics Division*
90, 39-68.

Kiff, P.R., 1973, Particle size analysis of sediments,
Laboratory Practice, April 1973, 259-266.

Kiff, P.R., 1977, The measurement of suspended solids with the
Partech turbidity monitor, *Equipment Notes No. 5,
Hydraulic Research Station,* Wallingford, England.

Lane, E.W., *et al.*,1947, Report of the sub-committee on sediment terminology, *Transactions, American Geophysical Union,* 28, 936-938.

Lane, E.W., and Borland, W.M., 1951, Estimating the bed load *Transactions, American Geophysical Union,* 32, 121-123.

Leopold, L.B., and Emmett, W.W., 1976, Bed-load measurements, East Fork River, Wyoming, *National Academy of Science Proceedings,* 73, 1000-1004.

Leopold, L.A., Wolman, M.G., and Miller, J.P., 1964, *Fluvial processes in geomorphology,* (Freeman and Co., San Francisco)

McHenry, J.R., *et al.*, 1967, Performance of nuclear sediment concentration gauges, in *Isotopes in hydrology, Proceedings of IAEA/IUGG Symposium, 1966,* (International Atomic Energy Agency, Vienna) 207-225.

Meyer-Peter, E., Favre, H., and Einstein, A., 1934, Neuere vemuchsresultate uber den Geschiebetrieb, *Schweiz. Bauzeitung,* 103, No. 13.

Murphree, C.E. *et al.*, 1968, Field test on x-ray sediment concentration guage, *Proceedings, American Society of Civil Engineers, Journal of Hydraulics Division,* 94, 515-528.

Nordin, C.F., 1977, Discussion of Yang, C.T., *et al.*, Applicability of unit stream power equation, *Proceedings, American Society of Civil Engineers, Journal of the Hydraulics Division,* 103, 209-211.

Serr, E.F., 1951, Measurement of bed-load sediment, *Transactions, American Geophysical Union,* 32, 123-126.

Sheldon, R.W., and Parsons, T.R., 1967, *A practical manual on the use of the Coulter counter in marine science,* (Coulter Electronics, Canada).

Smith, D.B., and Parsons, T.V., 1967, Radioisotope techniques for determining silt movement from spoil grounds in the Firth of Forth, in *Isotopes in Hydrology, Proceedings of IAEA/IUGG Symposium, 1966,* International Atomic Energy Agency, Vienna, 167-180.

Temple, P.H., and Sundborg, a., 1972, *The Rufiji River, Tanzania. Hydrology and sediment Transport,* Geografiska Annaler, Vol 54, Ser 9A, pp.345-368

Thorn, M.F.C., 1975, Monitoring silt movement in suspension in a tidal estuary, *16th Congress, International Association for Hydraulic Research, San Paulo, Brazil.*

Thorn, M.F.C., and Burt, T.N., 1977, Transport of Suspended sediment in the tidal River Crouch, *Report No. INT 148, Hydraulics Research Station, Wallingford, England.*

US Interagency Committee on Water Resources, 1963, A study of methods used in measurement and analysis of sediment loads in streams, *Report No. 14,* (Minneapolis, Minnesota).

Vanoni, V.A. (ed), 1975, *Sedimentation engineering, American Society of Civil Engineers, Manuals and Reports on Engineering Practise* No. 54.

Ward, P.R.B., 1978, Sedimnet yields from Rhodesian rivers 1976-1977 season, *Internal Report, Hydrological Branch, Ministry of Water Development,* (Salisbury, Rhodesia).

Ward, P.R.B., 1980, Sediment Transport and a Reservoir Siltation Formula for Zimbabwe, *The Civil Engineer in South Africa,* January 1980, pp. 9-16.

Water Resources Council, Sedimentation Committee, 1976, A study of methods used in movement and analysis of sediment loads in streams, *Catalog, Instruments and reports for fluvial sediment investigations* (revised edition).

Wertheimer, A.L., and Wilcock, W.L., 1976, Light scattering measurements of particle distribution, *Applied Optics* 15, 1616-1620.

4.

Sediment yield modelling

C. A. Onstad

Introduction

Sediment yield is "the total sediment outflow from a catchment or drainage basin, measurable at a point of reference and a specified period of time" (ASCE, 1975). It is the portion of the gross erosion within a basin that is not deposited before being transported from the basin. Sediment sources include upland sheet-rill erosion, gullies, streambanks, channels, construction sites, spoil banks, and roadsides. The relative magnitude of these potential sources depends on factors that include slope steepness and length, slope shape, soil type, land use, and rainfall characteristics.

Bennett (1974) divided the sediment yield process into the upland phase and the lowland phase. Sediment detachment and transport which depend on individual storm events predominate in the upland phase. Gully and sheet erosion, landslides, and construction sites are included in this category. Sediment yield from sheet-rill sources is usually greater than yields from other sources (Glymph, 1951: Leopold *et al.*,1966) Piest *et al.* (1975) reported that about 20% of the total sediment yield from a field-sized watershed resulted from gully erosion.

In the lowland phase, sediment transport and deposition processes predominate, and channel transport capacity becomes a significant factor. Rates of sediment transport in channels are determined by hydraulic variables, such as flow depth and velocity, slope, and the physical properties of sediment particles. In this phase, the effects of individual storm events are highly lagged and damped.

McGuinness *et al.* (1971) found that sediment yield from small basins in Ohio correlated well with rainfall characteristics and vegetative cover factors, whereas yield from a large basin correlated well with river flows. Otterby and Onstad (1981) found similar trends for catchments in Minnesota ranging in size from 362 km^2 to 38 600 km^2. Much of the sediment removed from upland basins is deposited where it may be mobilized again by lowland process rates.

Agricultural Engineer, USDA-ARS, Morris, MN 56267, USA.

The purpose of this chapter is to describe some specific sediment yield prediction needs and to describe some methods for predicting sediment yields.

Prediction needs

Specific needs for sediment yield prediction are so varied that no single model could meet them all without great loss of efficiency. Sediment problems can be divided into three broad categories (Williams, 1981): (1) erosion control planning, (2) water resources planning and design, and (3) water quality modelling. Some of these problems can be solved with simple models, but others require complex models. However, prediction requirements for each of these problems are largely determined by duration of the event to be simulated, size of area to be simulated and whether or not sediment sources are required (Onstad *et al.*, 1977). Not considered here is modelling for economic studies.

Time base

For solving a sediment yield problem or for designing a model of sediment yield, it is important to determine the appropriate time base. For situations where animals and plants are affected by high concentrations of sediments or chemicals, storm-based modelling of concentrations is required. Peak sediment concentrations over a specified time period are usually needed for determining compliance with water quality standards. A design storm simulation may be required for some conservation practices where quantitites of sediment are needed. Consideration of adsorbed chemicals may require information on variations in sediment-associated chemical concentrations with time in addition to sediment quantities.

For other problems, longer simulation periods may be more useful. For example, seasonal or monthly variations would be necessary to determine differences between winter cover and growing season conditions on upland areas. Such determinations are required for selection of land use and soil management techniques to control sediment yield and runoff.

Average annual sediment yield estimates are sufficient for design needs of reservoirs and conservation structures or for other studies of long-term sediment deposition. Structures for ponding water are built to store sediment for some time before filling. Average annual sediment yield estimates determine the amount of storage necessary and consequently the economic feasibility of the structure.

Channel maintenance plans require the prediction of long-term sediment yield trends. Studies of dredging to maintain drainage and navigation generally require long term estimates of sediment yield which unlike average annual or annual sediment yields indicate trends.

Basin size

Large basins usually require less modelling detail than
small basins. In large basins, channel definition is more
evident and more sediment sources of different types may be
active. As a result, modelling needs will usually dictate
the type of predictors necessary for various basin sizes.
Larger basins tend to behave more homogeneously than
small ones in a given physiographic region due to the
integration of many local processes. The groundwater
component of total runoff is usually higher for a large
basin than for a small one. Sediment sources are also more
varied. For example, gully and channel sediment contributions
must be considered in addition to rill-sheet contributions.
Bed load contributions generally also become more important
for large basins.
The outstanding feature of small basins is their
variability. This can result for a variety of reasons,
including topography, land use, cultural practices, soil
type, channel characteristics, and outlet conditions. Models
for predicting sediment yields under such varying conditions
must provide input parameters that are sensitive to the
important varying conditions.
For certain problems, specific parts of a basin will be
investigated. For example, the flood plain of a basin must
be investigated to establish zoning requirements. Another
example would be assessment of the sediment contributions
from various upland soil type-slope steepness associations
or from construction sites.

Sediment source

The number and types of sediment sources within a basin
and their relative importance, dictate modelling requirements.
Gullies contribute large amounts of sediment per unit area.
Such sediment is readily available for transport, because
gullies are generally located at the heads of channel
systems.
Urban areas within a basin pose a special problem with
their high runoff rates and transport capacities. Sediment
contributions from roadsides and ditches are often
appreciable. Channels, particularly in larger basins, may
contribute significantly to sediment yield. Conservation
structures present a problem for predicting yields from
their contributing areas. Construction sites with their
potential for extreme erosion rates also create simulation
problems related to their position within the catchment.
Sediment particle size distribution is important for
several reasons. Sand is transported differently from silt
and clay and presents a special computational need. Sand
sizes, generally greater than 0.062 mm, travel primarily as
bed load in channels. Likewise, sand is more easily
deposited near erosion sites in a catchment. If the
sediment is predominantly silt and clay (fines), a single
relation may be used to estimate sediment transport.   A
combination of sands and fine particles require a more
complex modelling procedure.

The sand fraction of sediment yield merits special
attention because it usually causes the most damage upon
deposition. Models must consider that sand beds will
fluidize at high discharges, changing the flow cross-section
and hence changing transport characteristics. If channels
become clogged with sand, transport capacity is reduced and
overbank flow will result, causing damage to flood plains.

Sediment yield prediction methods

Many sediment yield prediction methods are available and
have been used for various purposes. These methods can
generally be grouped into five categories: (1) sediment
delivery ratio procedures, (2) sediment rating curves,
(3) statistical equations, (4) deterministic models including
empirical parametric approaches and those using time-variant
interactions of physical processes and (5) stochastic
approaches.

Sediment delivery ratio methods

The sediment delivery ratio (SDR) method of predicting
sediment yield is well known and is used regularly
throughout the world. A SDR is defined as the fraction of
gross erosion (EROS) that is transported from a given basin
as sediment yield (SY).

$$SDR = SY/EROS \tag{1}$$

Gross erosion is composed of total sheet, rill, channel, and
gully erosion within a basin. The Universal Soil Loss
Equation (USLE) (Wischmeier and Smith, 1978) is currently
the most accepted method for predicting sheet and rill
erosion. The Soil Conservation Service (1966, 1971) has
developed guidelines for estimating gully and channel erosion.
Applying a delivery ratio to estimated gross erosion can
be a fairly accurate technique of predicting downstream
sediment yields if delivery ratios can be predicted accurately.
Delivery ratios have been estimated by comparing measured
sediment yield data with predicted gross erosion. These
delivery ratios have been related to basin and climate
characteristics to develop delivery ratio prediction
equations for ungauged basins.

Maner (1962) derived a relationship

$$\log SDR = 1.8768 - 0.14191 \log (10A) \tag{2}$$

where A = basin area

for the Blackland Prairie in Texas. Maner (1958) also
developed a relationship between sediment delivery ratio and
catchment relief-length ratio (R/L).

$$\log SDR = 2.94257 - 0.82363 \operatorname{colog} (R/L) \tag{3}$$

This relationship applies to the Red Hills physiographic area
of Oklahoma and Texas. Other relations have been developed
for use on ungauged basins (Gottschalk and Brune, 1950; Roehl,
1962; Williams and Berndt, 1972).

74

Williams (1977) has developed a procedure for
determining delivery ratios using sediment and runoff models,
thus eliminating long-term data collection. He demonstrated
the procedure on 15 Texas watersheds where it explained about
80% of the variation in average annual sediment yield.

Sediment rating curves

The sediment rating curve (Campbell and Bauder, 1940)
expressing the relationship between water discharge and
sediment discharge rate can be constructed by sampling
streamflow. Sediment yield frequency distributions can be
established using flow frequency distributions and sediment
rating curves (Williams, 1974). This method of estimating
sediment yield is time consuming and costly. Furthermore,
changing land uses and management practices may prevent
the establishment of a reliable relationship.
    Calibrated sediment transport equations can be used to
calculate sediment rating curves. Such equations have been
developed by Einstein (1950) Colby and Hembree (1955) and
Laursen (1958). More recently, equations by Ackers and White
(1973), Engelund and Hansen (1967), and Yang (1973) appear
promising.

Statistical equations

Sediment yield equations derived from statistical
analyses have been frequently used to estimate sediment
yield. These types of equations usually relate sediment
yield to one or more basin characteristics or climatic
factors. Because of their nature, they require relatively
large quantities of data both on basin characteristics and
on sediment discharge.
    Statistical analysis is commonly used for problems
requiring sediment yield averages over long time periods.
Generally, basin areas studied are relatively large and
include basins used for water supplies.
    Flaxman (1972) developed and revised in 1974 a regression
equation principally for reservoir design on rangeland basins
in the western USA. Flaxman's (1974) expression is:

$$\log (Y+100) = 524.2 - 270.7 \log(X1+100) + 6.4 \log(X2+100)-$$
$$- 1.7 \log(X3+100) + 4.0 \log(X4+100) + \log(X5+100)(4)$$

where Y = average annual sediment yield (t/mi^2)

    X1 = ratio of average annual precipitation (in) to
        average annual temperature $(^{\circ}F)$

    X2 = average basin slope (%)

    X3 = soil particles greater than 1.0 mm (%)

    X4 = soil aggregation index (%)

    X5 = 50% probability peak discharge $(ft^3/s/mi^2)$

These five parameters are expressions of vegetative growth
(X1), topography (X2), soil properties (X3 and X4), and
climate (X5). This equation explained about 91% of the
variance in average sediment yield from 27 basins ranging in
size from 12 to 54 mi^2 in 10 western states of the USA.

Anderson (1976) used data from 48 forested basins in northern California to derive an expression for sediment yield based on 34 independent variables. The variables included topography, geology, roads, fires, hydrology and soil. The equation is used for evaluating the effects of forest fires and road development on sediment deposition.

Other researchers have developed similar equations for other localized areas. Branson and Owen (1970) used geomorphic variables, watershed cover, and hydrologic variables to develop a sediment yield equation for a basin in western Colorado. Other notable examples are the work of Tatum (1963) for the Pacific slopes of California, Hinda (1976) for basins in Wisconsin, and Herb and Yorke (1976) f construction sites in the Washington, DC area.

Dendy and Bolton (1976) derived sediment yield equation from reservoir deposition data obtained from throughout the USA. They related deposition in about 800 reservoirs to drainage area and mean annual runoff for basins ranging in size from 1 mi^2 (2.59 km^2) to 30 000 mi^2 (77 700 km^2) with runoff ranging from near zero to 50 in/year (1270 mm/year). For areas where runoff is less than 2 in/year (50 mm/year), their derived equation is:

$$S = 1280 \ Q^{0.46}(1.43 - 0.26 \ \log A) \tag{5}$$

and for other areas:

$$S = 1958 \ e^{-0.055Q} \ (1.43 - 0.26 \ \log A) \tag{6}$$

where

S = sediment yield (t/mi^2/year)

Q = runoff (in)

A = basin area (mi^2)

The coefficient of variation for the two equations is 0.75 t/mi^2/year. These equations express a good general relationship for sediment yield on a regional basis.

Wallis and Anderson (1965) utilized multivariate techniques including principle component analysis and principle components regression to develop a simplified prediction equation for sediment yield. They pointed out, however, that one of the limitations of statistical approac is that they cannot be used without recalibration where bas management changes have occurred due to man's activities.

Deterministic models

Deterministic models introduce parameters to quantify the factors affecting erosion, transport, and deposition. These parameters can be derived empirically or calibrated using curve fitting techniques. One example of a parametric model is the Wischmeier and Smith (1978) soil loss equation:

$$A = RKSLCP \tag{7}$$

where A = average annual soil loss (t/ha)

R = rainfall and runoff factor (EI, energy-
intensity units)

K = soil erodibility factor

S = slope steepness factor

L = slope length factor

C = cover and management factor

P = support practice factor

Many parametric sediment yield models for basins use this equation as a base because of the widespread availability of the parameter values in the USA.

Williams (1975) developed a model to predict storm sediment yield for basins up to 2600 km^2. His equation, termed the modified universal soil loss equation (MUSLE) is:

$$Y = 95(Qq_p)^{0.56}KCPLS \tag{8}$$

where Q = runoff volume (m^3)

q_p = peak runoff rate (m^3/s)

and Y = sediment yield (t)

The other factors are identical to those in equation (7). The variables Q and q_p are estimated, using a water yield model developed by Williams and LaSeur (1976).

Runoff and sediment are routed using hydraulic parameters of the basin, decay functions, and sediment characteristics. Average annual sediment yield is computed by integrating a frequency curve of storm sediment yields and dividing by the largest return period.

Other parametric models using modified versions of the USLE (Wischmeier and Smith, 1978) have incorporated the soil detachment-soil transport concept of Meyer and Wischmeier (1969). For most empirical parametric approaches the detachment phase is usually represented by some form of the USLE equation. However, the transport phase has been expressed in several ways with varying degrees of success. Such models are frequently called sediment routing models.

Sediment routing increases the accuracy of sediment yield predictions on large basins and allows determination of subbasin contributions to total sediment yield. In addition, sediment sources can be located and ranked within the basin. Sediment characteristics such as changes in particle-size distribution can also be considered in routing models.

Kling (1974) and Kling and Olson (1974) developed an annual sediment yield model using the detachment-transport concept for a basin in New York State. They subdivided a 117 mi^2 catchment into square 10 acre cells for purposes of calculation. Soil detachment for each cell was estimated using the USLE (Wischmeier and Smith, 1978). The drainage direction for each cell was determined to define the routing system. They also developed a transport factor, which indicated the efficiency of eroded material transport from cell to cell within the drainage network.

The transport factor was based on comparisons of slope steepnesses for adjacent cells. If the slope of a cell was greater than the slope of the adjacent upstream cell, it was assumed that all sediment was transported. If the cell slopes were conversely related, only a portion of the sediment was transported depending on the ratio of the slope steepnesses. This procedure produced fair results for a test basin in Pennyslavania.

Onstad and Foster (1975) developed a single-storm sediment yield model for small watersheds. They subdivided basins using the stream path analogy of Onstad and Brakensiek (1968). Using this analogy, contour lines are considered to be equipotential lines perpendicular to which flow lines are drawn. The cell, which is the basic unit for purposes of calculation, is bounded by adjacent flow lines and contour lines. All flow is assumed to move from cell to cell across the contour lines until the stream channel is reached. For this model, detachment is represented as follows after Foster and Wischmeier (1974):

$$E_j = W_j (KCPS)_j (x_j^{1.5} - x_{j-1}^{1.5}) \qquad (9)$$

where E_j = detachment capacity on segment j

x_j = distance from top of slope to lower end of segment j

w_j = energy term.

The energy term, W_j, of equation (9) is represented by a combination of rainfall and runoff energy as follows:

$$W_j = 0.5 R_{st} + 15 Qq_{pj}^{1/3} \qquad (10)$$

where R_{st} = storm EI factor (Wischmeier and Smith, 1978)

The cumulative detachment capacity for the entire slope is the sum of E_j over all the slope segments.

The transport phase of this model is also patterned after the Wischmeier and Smith (1978) equation. The transport capacity at any point on a slope is presented by

$$T_{cx} = WKSCP \; x^{1.5} \qquad (11)$$

where T_{cx} = transport capacity at location x.

W = transport energy term

The values for the parameters K, S, C, and P are taken to be those established by Wischmeier and Smith (1978), although they may be different between the transport and detachment phases. Niebling and Foster (1977) have proposed numerical values for these and other factors involved in the transport process.

Results of detailed testing of the Onstad and Foster (1975) model, the Williams (1975) model, and the USLE (Wischmeier and Smith, 1978) coupled with a sediment delivery ratio for two small catchments in Iowa is reported by Onstad and others (1976). They found that the Williams model performed best for the smaller sediment yield events and that the Onstad-Foster model performed best for the larger events and overall. For the 110 individual events tested, the

Onstad-Foster model predicted sediment yields with a standard error of 0.8 tons/acre.

Onstad and Bowie (1977) simplified the routing model to route average annual sediment yields. The simplified model neglected particle size and did not use a degradation component, but it is convenient and represents an improvement over gross erosion-delivery ratio methods (Williams, 1981). An early version of a sediment routing model by Williams (1975) routed sediment to the basin outlet based on travel time from subbasins and median particle size. In was refined (Williams and Hann, 1978) by replacing the median particle size with the entire particle-size distribution. Further refinement (Williams, 1978) included the development of degradation-component based on the concept of stream power.

Foster and Meyer (1975) developed a deterministic sediment yield based on the continuity equation for mass transport:

$$\frac{\delta G}{\delta x} = D_r + D_i \qquad (12)$$

where δG = sediment load

δx = distance downslope

D_r = detachment rate of rill erosion

and D_i = delivery rate of particles detached by interrill erosion to rill flow.

They also included an approximate expression for deposition:

$$\frac{D_r}{D_{cr}} + \frac{G}{T_{cr}} = 1 \qquad (13)$$

where D_{cr} = detachment capacity of rill flow

T_{cr} = transport capacity of rill flow.

The variables, D_r, T_{cr}, and D_i are functions of rainfall and runoff characteristics, soil properties, and vegetative cover. The two unknowns in equations (12) and (13) are D_r, the rill detachment rate, and G, the sediment load. Combining equations (12) and (13) by eliminating D_r produces:

$$\frac{\delta G}{\delta x} = \frac{D_{cr}}{T_{cr}} (T_{cr} - G) + D_i \qquad (14)$$

The rate of particle delivery detachment by interrill erosion, D_i, depends on rainfall energy and on the transport capacity of interrill flow. This in turn depends on rainfall intensity, rainfall excess, and rill geometry.

The detachment capacity of rill erosion, D_{cr}, and the transport capacity of rillflow, T_{cr}, are functions of bed shear, τ, and can be approximated by:

$$D_{cr} = C_c \tau \ 3/2 \qquad (15)$$

and

$$T_{cr} = C_t \tau \ 3/2 \qquad (16)$$

where C_c = coefficient of detachment capacity

C_t = coefficient of transport capacity.

79

The following expression is obtained when equations (15) and (16) are substituted into equation (14):

$$\frac{\delta G}{\delta x} = C_d (C_t \ \tau^{3/2} - G) + D \qquad (17)$$

where $C_d = C_c / C_t$.

Bed shear, for steady state conditions, is a function of rainfall excess rate, slope, and flow distance. Equation (17) has been expressed in dimensionless form and solved for uniform, concave, and complex slopes (Foster and Meyer, 1975

Shirley and Lane (1978) used the equations of rill and interrill erosion and transport with the kinematic flow equations to obtain analytic solutions for sediment concentrations throughout a hydrograph. Their equations are as follows:

$$\frac{\delta h}{\delta t} + \frac{\delta q}{\delta x} = R \qquad (18)$$

and
$$q = Kh^m \qquad (19)$$

where h = flow depth

t = time

q = runoff

x = distance

R = rainfall excess rate

K = slope-resistance coefficient

m = dimensionless exponent.

Later Lane and others (1982) derived separate equations for bed load transport and suspended load transport based on the assumption that sediment particles greater than 0.062 mm in diameter travel as bed load and that particles finer than 0.062 mm in diameter travel as suspended load. They describe transport capacity of bed load particles as:

$$g_{sb}(d_i) = \alpha \ f_i \ B_s(d_i) \ T(T - T_c(d_i)) \qquad (20)$$

where $g_{sb}(d_i)$ = transport capacity for particles of size d_i

α = a weighting factor to assure that the sums of individual transport capacities equals total transport capacity.

f_i = proportion of particles in size class i

d_i = median diameter of particles in size class i

$B_s(d_i)$ = sediment transport coefficient

T = shear stress

$T_c(d_i)$ = critical shear stress for particles for size i.

80

The total bed load transport capacity is found by summing the results of equation (20) over all size fractions.

For suspended load, Lane and Hakonson (1982) used the relations developed by Bagnold (1956, 1966) based on the concept of streampower:

$$i_s = P \frac{e_s u_s}{v_s} (1 - e_b) \tag{21}$$

where
i_s = suspended sediment transport capacity

$P = \tau V$ = available stream power per unit area of bed

e_s = suspended load efficiency factor

e_b = bed load efficiency factor

u_s = transport velocity of suspended load

v_s = settling velocity of the particles

τ = shear stress

V = flow velocity

If u_s is assumed equal to the mean fluid velocity, then equation (21) is of the form:

$$g_{sus} = k \tau V^2 \tag{22}$$

where the coefficient k includes the efficiency parameter, the settling velocities, and the proportion of particles smaller than 0.062 mm in the channel bed material. The total load is computed as the sum of the bed load, equation (20), and the suspended load, equation (22).

The model was used to predict sediment yields for 47 runoff events for five small catchments in Arizona and for 27 events on the Niobrara River in Nebraska. The coefficients of determination were 0.78 and 0.97, respectively, for predictions on the two catchments.

Other time variant deterministic models have been developed by Smith (1976); Foster, Meyer, and Onstad (1977); Johanson and Leytham (1977); Fleming and Walker (1977); and Bennett and Nordin (1977).

Stochastic models.

Only recently have stochastic models been considered for sediment yield predictions. Among the first models were those developed by Murota and Hashino (1969) and Woolhiser and Todorovic (1971).

Murota and Hashino (1969) used an approach that develops a deterministic relationship between sediment yield and daily rainfall and develops distribution functions for total sediment yield in an n-day period. They assumed a linear relation between runoff and daily rainfall. Knowing the relation between runoff and stage, they calculated sediment yield due to hydraulic forces on the channel. Contributions from upland erosion sources were ignored. Using their method, a threshold amount of rainfall was determined below which no sediment would move. They used Monte Carlo simulation and numerical integration and summation to obtain distributions of sediment yield and runoff for nine periods within a calendar year.

Renard and Lane (1975) and Renard and Laursen (1975) used stochastic simulation techniques to estimate distribution functions of sediment yield from semiarid watersheds. Their model consisted of a stochastic runoff model together with a deterministic equation to calculate sediment yield from a runoff hydrograph. The two variables used to describe the runoff season are the starting date and the number of runoff events. Time-of-day and the interval between events were used to describe the temporal distribution. Each runoff event was described by runoff volume and peak discharge. Manning's equation and Laursen' (1958) equation were used to calculate sediment transport rates at three or five points on the triangular hydrograph.

Fogel and others (1976) used two empirical expressions relating runoff to rainfall characteristics, and sediment yield to both rainfall and runoff characteristics. The runoff storm volume is calculated as follows based on the Soil Conservation Service (SCS) (1972):

$$Q = \frac{(R-A)^2}{R-A+S} \tag{23}$$

where Q = storm runoff volume

R = storm rainfall

A = initial abstraction

S = potential infiltration

The peak flow rate also based on the SCS (1971) is:

$$q_p = \frac{484\ A_w\ V}{0.5\ D + 0.6\ T_c} \tag{24}$$

where q_p = peak flow rate

A_w = catchment area

D = duration of the rainfall excess

T_c = time of concentration.

For sediment yield they used the work of Williams (1975) expressed in equation (8).

Substituting equations (23) and (24) into equation (8) results in the following expression for sediment yield per storm, Z, in terms of several catchment characteristics and two rainfall variables R and D:

$$Z = 95 \left[\frac{484\ A_w (R-A)^4}{(0.5D + 0.6T_c)\ (R-A+S)^2} \right]^{0.56} KCPLS \tag{25}$$

In equation (25), R and D were assumed to be distributed as bivariate gamma distribution. The total seasonal sediment yield could be calculated by assuming that the probability an event of significance is very small.

Simons and others (1976) developed a method using Williams (1975) equation for suspended sediment yield. They used a method developed by Reich (1962) to determine runoff volume and a method detailed by Reich and Heimstra (1965) to predict peak flows. These procedures use basin, storm and soil characteristics in their equations. For bed load, they used the equation developed by Meyer-Peter and Muller (US Bureau of Reclamation, 1960). They used the concept of return-periods to estimate sediment yield for several different conditions of data availability ranging from no data for either rainfall or flow to complete data on both.

Woolhiser and Blinco (1975) developed several general models treating sediment yield as a stochastic process. They included procedures to be used for both small and large catchments and demonstrated possible advantages of the stochastic approach over methods presently in use. The models consist of procedures to determine detachment, entrainment, transport, and deposition of sediment. The hypotheses were not tested with actual field data.

Summary

Sediment yield models vary considerably to accomodate the wide range of water resources problems where sediment yield estimates are needed. Williams (1981) has categorized these problems as follows: (1) erosion control planning, (2) water resources planning and design, and (3) water quality modelling. The simplest models are generally used in erosion control planning where only the mean sediment yield is needed. However, more complex models have been developed and used to estimate sediment yields from individual storms. Certain water resource planning activities and water quality modelling usually requires simulation. Here the time sequence of sediment yield is included. This requires a more general formulation involving time distributions of sediment producing rainstorms, runoff events, or of the sediment yield events themselves. The selection of the proper model to be used for a particular problem depends on many factors including the available data and the risk involved.

REFERENCES

Ackers, P., and White, W.R., 1973, Sediment transport: New approach and analysis, *Proceedings, American Society of Civil Engineers, Journal of the Hydraulics Division,* 99, 2041-2060.

American Society of Civil Engineers, 1975, *Sedimentation Engineering, American Society of Civil Engineers Manuals and Reports on Engineering Practice,* No. 54.

Anderson, H.W., 1976, Reservoir sedimentation associated with catchment attributes, landslide potential, geologic faults, and soil characteristics. *Proceedings, Third Federal Inter-Agency Sedimentation Conference,* 1.35-1.46.

Bagnold, R.A., 1956, The flow of cohesionless grains in fluids. *Philosophical Transactions of the Royal Society of London*, Series A, 246(964), 235-297.

Bagnold, R.A., 1966, An approach to the sediment transport problem from general physics. *US Geological Survey Professional Paper*, 422-I.

Bennett, J.P., 1974, Concepts of mathematical modeling of sediment yield. *Water Resources Research*, 10, 485-492.

Bennett, J.P., and Nordin, C.F., 1977, Simulation of sedimen transport and armoring. *Hydrological Sciences Bulletin*, 22, 555-569.

Branson, F.A., and Owen, J.R., 1970, Plant cover, runoff, an sediment yield relationships on Mancos shale in western Colorado. *Water Resources Research*, 6, 783-790.

Campbell, F.B., and Bauder, H.A., 1940, A rating curve metho for determining silt discharge of streams. *Transactions, American Geophysical Union*, 21, 603-60

Colby, B.R., and Hembree, C.H., 1955, Computations of total sediment discharge, Niobrara River near Cody, Nebraska. *US Geological Survey Water Supply Paper*, 1357.

Dendy, F.E., and Bolton, G.C., 1976, Sediment yield-runoff-drainage area relationships in the United States, *Journal of Soil and Water Conservation*, 31, 264-266.

Einstein, H.A., 1950, The bedload function for sediment transportation in open channel flows. *US Department of Agriculture Technical Bulletin* No. 1026.

Engelund, F., and Hansen, E., 1967, A monograph on sediment transport in alluvial streams, *Teknish Vorlag*, Copenhagen.

Flaxman, E.M., 1972, Predicting sediment yield in western United States. *Proceedings, American Society of Civil Engineers, Journal of the Hydraulics Division*, 98, 2073-2085.

Flaxman, E.M., 1974, Predicting sediment yield in western United States, in *Pacific Southwest Inter-Agency Committee, Report of Water Management Subcommittee, Erosion and Sediment Yield Methods*.

Fleming, G., and Walker, R.A., 1977, Digital simulation of soil erosion from land surface, in *Proceedings, International Association of Hydraulic Research Symposium of Baden-Baden*.

Fogel, M.M, Hekman, L.H., and Duckstein, L., 1976, A stochastic sediment yield model using the modified universal soil loss equation, in *Soil erosion: Prediction and control*, *(Soil Conservation Society of America, Ankery, Iowa)* 226-233.

Foster, G.R., and Meyer, L.D., 1975, Mathematical simulation of upland erosion by fundamental erosion mechanics, in, *Present and prospective technology for predicting sediment yields and sources*, US Department of Agriculture, Agricultural Research Service Publication No. ARS-S-40, 190-207.

Foster, G.R., Meyer, L.D., and Onstad, C.A., 1977, An erosion equation derived from basin erosion principles, *Transactions, American Society of Agricultural Engineers*, 20, 678-682.

Foster, G.R., and Wischmeier, W.H., 1974, Evaluating irregular slopes for soil loss predictions, *Transactions, American Society of Agricultural Engineers*, 17, 305-309.

Gottschalk, L.C., and Brune, G.M., 1950, Sediment design criteria for the Missouri Basin loess hills, *US Department of Agriculture, Soil Conservation Service, Technical Paper* No. 97.

Glymph, L.M., 1951, Relation of sedimentation to accelerated erosion in the Missouri River Basin, *US Department of Agriculture, Soil Conservation Service, Technical Paper* No. 102.

Herb, W.J., and Yorke, T.H., 1976, Storm period variables affecting sediment transport from urban construction areas. *Proceedings, Third Federal Inter-Agency Sedimentation Conference*, 1.181-1.192.

Hindall, S.M., 1976, Prediction of sediment yields in Wisconsin streams. *Proceedings, Third Federal Inter-Agency Sedimentation Conference*, 1.205-1.218.

Johanson, R.C., and Leytham, K.M., 1977, Modelling sediment transport in natural channels, in *Proceedings of the Watershed Research Workshop*, (Smithsonian Institute, Edgewater, Maryland).

Kling, G.F., 1974, A computer model of diffuse sources of sediment and phosphorus moving into a lake, unpublished Ph.D. Thesis, Cornell University, Ithaca, New York.

Kling, G.F., and Olson, G.W., 1974, Computer modelling of sediment and phosphorus movement into Candarago Lake. *Cornell University, Department of Agronomy, Ithaca, New York, Mimeo Report* No. 74-11.

Lane, L.J., and Hakonson, T.E., 1982, Influence of particle
    sorting in transport of sediment associated
    contaminants, in *Proceedings of the Tucson Waste
    Management Symposium.*

Lane, L.J., White, G.C., and Hakonson, T.E., 1982, Simulati
    of sediment transport in alluvial channels and spec
    applications, *Los Alamos National Laboratory Report.*

Laursen, E.M., 1958, The total sediment load of streams.
    *Proceedings, American Society of Civil Engineers,
    Journal of the Hydraulics Division*, 84, 1530-1536.

Leopold, L.B., Emmett, W.W., and Myrick, R.M., 1966, Channe
    and hillslope processes in a semiarid area in New
    Mexico, *US Geological Survey Professional Paper* 352

McGuiness, J.L., Harrold, L.L., and Edwards, W.M., 1971,
    Relation of rainfall energy and streamflow to sedim
    yield from small and large watersheds. *Journal of
    Soil and Water Conservation*, 26, 233-234.

Maner, S.B., 1958, Factors affecting sediment delivery rate
    in the Red Hills physiographic area. *Transactions,
    American Geophysical Union*, 39, 669-675.

Maner, S.B., 1962, Factors influencing sediment delivery
    ratios in the Blackland Prairie land resource area,
    *US Department of Agriculture, Soil Conservation
    Service Report*, Fort Worth, Texas.

Meyer, L.D., and Wischmeier, W.H., 1969, Mathematical
    simulation of the process of soil erosion by water.
    *Transactions, American Society of Agricultural
    Engineers*, 12, 754-759.

Murota, A., and Hashino,M., 1969, Studies of a stochastic
    rainfall model and its application to sediment
    transportation, *Technical Reports of the Osaka
    University*, 19, 231-247.

Neibling, W.H., and foster, G.R., 1977, Estimating depositi
    and sediment yield from overland flow processes, in
    *Proceedings of the International symposium on Urban
    Hydrology, Hydraulics, and Sediment Control*,
    University of Kentucky, Lexington, 75-86.

Onstad, C.A., and Bowie, A.J., 1977, Watershed sediment yie
    modeling using hydrolic variables, in *Proceedings
    of the International Symposium on Erosion and Solid
    Matter Transport in Inland Waters*, International
    Association of Hydrological sciences Publication No
    122, 191-202.

Onstad, C.A., and Brakensiek, D.L., 1968, Watershed simulat
    by stream path analogy, *Water Resources Research*, 4
    965-972.

Onstad, C.A., and Foster, G.R., 1975, Erosion modeling on a watershed. *Transactions, American Society of Agricultural Engineers*, 18, 288-292.

Onstad, C.A., Mutchler, C.K., and Bowie, A.J., 1977, Predicting sediment yields, in *Soil Erosion and Sedimentation, Proceedings of the National Symposium on Soil Erosion and Sedimentation*, American Society of Agricultural Engineers Publication No. 4-77, 43-58.

Onstad, C.A., Piest, R.F., and Saxton, K.E., 1976, Watershed erosion model variation for Southwest Iowa, *Proceedings, Third Federal Inter-Agency Sedimentation Conference*, 1.22-1.34.

Otterby, M.A., and Onstad, C.A., 1981, Average annual sediment yields in Minnesota, US Department of Agriculture, Agricultural Research Service Report No. ARR-NC-8.

Piest, R.F., Kramer, L.A., and Heinemann, H.G., 1975, Sediment movement from loessial watersheds, in *Present and prospective technology for predicting sediment yields and sources*, US Department of Agriculture, Agricultural Research Service Publication No. ARS-S-40, 130-141.

Reich, B.M., 1962, Design hydrographs for very small watersheds from rainfall, *Ph.D. Dissertation, Colorado State University, Fort Collins*.

Reich, B.M., and Heimstra, A.V., 1965, Tacitly maximized small watershed flood estimates. *Proceedings, American Society of Civil Engineers, Journal of the Hydraulics Division*, 91, 217-245.

Renard, K.G., and Lane, L.J., 1975, Sediment yield as related to a stochastic model of ephemeral runoff, in *Present and prospective technology for predicting sediment yields and sources*, US Department of Agriculture, Agricultural Research Service, Publication No. ARS-S-40, 253 263.

Renard, K.G., and Laursen, E.M., 1975, A dynamic behavior model of an ephemeral stream, *Proceedings, American Society of Civil Engineers, Journal of the Hydraulics Division*, 101, 511-528.

Roehl, J.W., 1962, Sediment source areas, delivery ratios and influencing morphological factors, *International Association of Scientific Hydrology Publication*, No. 59, 202-213.

Shirley, E.D., and Lane, L.J., 1978, A sediment yield equation from an erosion simulation model, in *Hydraulics and Water Resources in Arizona and the Southwest, Proceedings, 1978 Meetings, Flagstaff*, (Arizona Academy of Science).

Simons, D.B., Reese, A.J., Li, R.M., and Ward, T.J., 1976, A simple method for estimating sediment yield. *Soil erosion: Prediction and control,* (Soil Conservation Society of America, Ankeny, Iowa), 234-241.

Smith, R.E., 1976, Simulating erosion dynamics with a deterministic distribution watershed model. *Proceedings, Third Federal Inter-Agency Sedimentati∢ Conference,* 1.163-1.173.

Soil Conservation Service, 1966, Procedures for determining rates of land damage, land depreciation, and volume of sediment produced by gully erosion, *US Departmen of Agriculture Technical Release* No. 32, Geology.

Soil Conservation Service, 1971, Sediment sources, yields, and delivery ratios, in *National Engineering Handbo∢* Section 3, Sedimentation, Chapter 6, (US Department of Agriculture, Washington, DC).

Soil Conservation Service, 1972, Hydrology, in *National Engineering Handbook,* (US Department of Agriculture Washington, DC).

Tatum, F.E., 1963, A new method of estimating debris-storag∢ requirements for debris basins, in *US Department of Agriculture Miscellaneous Publication,* No. 970, 886-898.

U.S. Bureau of Reclamation, 1960, *Investigation of the Meyer-Peter, Muller bedload formula,* (US Bureau of Reclamation, Denver).

Wallis, J.R., and Anderson, H.W., 1965, An application of multivariate analysis to sediment network design, *International Association of Scientific Hydrology Publication,* No. 67, 357-378.

Williams, J.R., 1974, Predicting sediment yield frequency f∢ rural basins to determine man's effect on long-term sedimentation, in *Effects of Man on the Interface o∂ the Hydrological Cycle with the Physical Environmen∖* International Association of Hydrological Sciences Publication No. 113, 105-108.

Williams, J.R., 1975, Sediment yield prediction with univers equation using runoff energy factor, in *Present and prospective technology for predicting sediment yiel∢ and sources,* US Department of Agriculture, Agricultural Research Service Publication No. ARS-S-40, 244-252.

Williams, J.R., 1977, Sediment delivery ratios determined with sediment and runoff models, in *Erosion and Sol∢ Matter Transport in Inland Waters,* International Association of Hydrological Sciences Publication No∢ 122, 168-179.

Williams, J.R., 1978, A sediment yield routing model, in *Proceedings of the Specialty Conference on Verification of Mathematical and Physical Models in Hydraulic Engineering, ASCE, College Park, MD.,* (American Society of Civil Engineers), 662-670.

Williams, J.R., 1981, Mathematical modeling of watershed sediment yield, in *Proceedings, International Symposium on Rainfall-Runoff Modeling, Mississippi State University.*

Williams, J.R., and Berndt, H.D., 1972, Sediment yield computed with universal equation, *Proceedings, American Society of Civil Engineers, Journal of the Hydraulics Division,* 98, 2087-2098.

Williams, J.R., and Hann, R.W., 1978, Optimal operation of large agricultural watersheds with water quality constraints, *Texas A & M University Technical Report* No. 96.

Williams, J.R., and LaSeur, W.V., 1976, Water yield model using SCS curve numbers, *Proceedings, American Society of Civil Engineers, Journal of the Hydraulics Division,* 102, 1241-1253.

Wischmeier, W.H., and Smith, D.D., 1978, Predicting rainfall -erosion losses - Guide to conservation planning, *US Department of Agriculture, Agriculture Handbook,* No. 537.

Woolhiser, D.A., and Blinco, P.H., 1975, Watershed sediment yield - a stochastic approach, in *Present and prospective technology for predicting sediment yield and sources, US Department of Agriculture, Agricultural Research Service Publication,* No. ARS-S-40, 264-273.

Woolhiser, D.A., and Todorovic, P., 1971, A stochastic model of sediment yield for ephemeral streams, in *Statistical Hydrology,* US Department of Agriculture, Miscellaneous Publication No. 1275.

Yang, C.T., 1973, Incipient motion and sediment transport, *Proceedings, American Society of Civil Engineers, Journal of the Hydraulics Division,* 99, 1679-1704.

5.

Measurement of bedload
in rivers

W. W. Emmett

Introduction

Schoklitsch (1950), in reference to bed load transport,
stated "...... there is not too much known about it." His
statement was not without reason; the problems associated
with measurement of bed load transport in alluvial channels
are significant.

Bed load

Bed load is that sediment carried down a river by
rolling and saltation on or near the streambed. Though bed
load may best be defined as that part of the sediment load
supported by frequent solid contact with the unmoving bed,
in practice it is the sediment moving on or near the
streambed rather than in the bulk of the flowing water.
In the sediment-transport process, individual bed-
material particles are lifted from the streambed and set
into motion. If the motion includes frequent contact of a
particle with the streambed, the particle constitutes part
of the bed load. If the motion includes no contact with the
streambed, the particle is literally a part of the suspended
load, regardless of how close to the streambed the motion
occurs. Depending on the hydraulics of flow in various
reaches of a channel, particles may alternate between being
a part of the bed load or a part of the suspended load. At
a given cross section of channel, particles that are part
of the bed load at one stage may be part of the suspended
load at another stage. Any particle in motion may come to
rest; for bed load, downstream progress is likely to be a
succession of movements and rest periods. Particles at rest
are part of the bed material. Obviously, there is an
intimate relation between bed material, bed load, and
suspended load.

Hydrologist (Engineer), U.S. Geological Survey, Water
Resources Division, Box 25046, Mail Stop 413, Denver
Federal Center, Lakewood, Colorado 80225, USA.

Owing to the somewhat nebulous definition of bed load, it becomes an exceedingly difficult task to build measuring equipment which samples only bed load. Any device which rests on the streambed is perilously close to sampling bed material, and any device which protrudes upward from the streambed, or be necessity is raised or lowered through the flow, may sample some part of the suspended load.

Hubbell (1964) has described many of the problems encountered with measurement of bed load and also provided a state-of-the-art report on apparatus and techniques for measuring bed load. This report presents information on the Helley-Smith bed load sampler, developed since the Hubbell report.

Helley-Smith bed load sampler

Helley and Smith (1971) introduced a pressure-difference bed load sampler that is a structurally modified version of the Arnhem sampler (Hubbell, 1964). The Helley-Smith bed load sampler has an expanding nozzle, sample bag, and frame (Figs 5.1 and 5.2). The sampler was designed to be used in flows with mean velocities to 3 m and sediment sizes from 2 to 10 m/s. The sampler has a square 7.62 cm entrance nozzle and a 46 cm-long sample bag constructed of 0.2 mm mesh polyester, though more recently it has become standard practice to use a sample bag of 0.25 mm mesh polyester. The standard sample bag has a surface area of approximately 1900 cm^2.

The original design included a brass nozzle, aluminum-tubing frame weighted with poured molten lead to a total weight of 30 kg, aluminum tail fins, and bolted construction. More recent versions of the sampler have stainless-steel nozzles for greater durability, steel-plate tail fins, solid-steel round-stock bar frame selected to maintain a 30 kg total weight, and all-welded construction. The sample bag attaches to the rear of the nozzle with a rubber "O" ring. A sliding bracket on the top frame member allows for cable-suspended lowering and raising of the sampler. Position of the bracket along the frame controls the sample attitude; normal attitude is a slightly tail-heavy position (about a 15 degree angle).

An extensively used version of the sampler has the nozzle and sample bag adapted to a wading rod, rather than having a frame and tail-fin assembly. To minimize weight and facilitate use of this model, the nozzle is generally of cast aluminum and equipped with a sectionalized tubular aluminum wading rod.

Calibration of the bed load sampler

Calibration refers to determining the hydraulic and sampling efficiencies of the sampler.

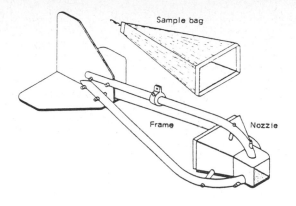

Sample bag

Frame   Nozzle

All Dimensions In Centimeters

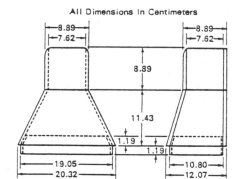

A. Plan and side elevation of 7.62 cm Helley-Smith
bedload sampler nozzle.

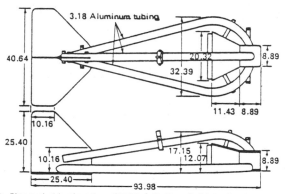

3.18 Aluminum tubing

B. Plan and side elevation of 7.62 cm Helley-Smith bedload sampler.

Fig 5.1.   The Helley-Smith bed load sampler.

Fig 5.2.   Plan and side elevation drawings of the Helley-smith bed load sampler.

Hydraulic characteristics

Hydraulic efficiency of a bed load sampler has been defined (Hubbell, 1964) as the ratio of the mean velocity of water discharge through the sampler to the mean velocity of the water discharge which would have occurred through the area occupied by the opening in the sampler nozzle had the sampler not been there.

A laboratory hydraulic calibration of the Helley-Smith bed load sampler was conducted at the US Geological Survey Gulf Coast Hydroscience Center (Druffel et. al., 1976). In the laboratory study, velocity profiles were measured in the sampler nozzle and at various locations upstream from the sampler. The results of this study showed that the hydraulic efficiency of the Helley-Smith bed load sampler is approximately 1.54. This value of hydraulic efficiency was found to be constant for the range of flow conditions in the experiments, a range applicable to many natural streamflow conditions.

The study, along with field observations by the writer, indicates the sample bag can be filled to 40% capacity with sediment larger than the mesh size (0.2-0.25 mm) of the bag without reduction in hydraulic efficiency. However, sedime with diameters close to the mesh size of the sample bag bot plugs the sample bag and escapes through the mesh, causing an unpredictable decrease in hydraulic efficiency and loss of the sample.

Data on hydraulic characteristics of the sampler provid qualitative information about probable performance, but suc data cannot be used directly to evaluate sediment-trap efficiency of the unit.

Sediment-trapping characteristics

The sampling efficiency of a bed load sampler is define (Hubbell, 1964) as the ratio of the weight of bed load collected during a sampling time to the weight of bed load that would have passed through the sampler width in the sam time, had the sampler not been there. Ideally, the ratio is 1.0, and the weight of every particle-size fraction in the collected sample is in the same proportion as in the true bed load discharge.

A field calibration of the sediment-trapping characteristics of the Helley-Smith bed load sampler was conducted at the US Geological Survey Bed Load Transport Research Facility on the East Fork River, Wyoming (Emmett, 1979). An open slot across the streambed of the East Fork River, continually evacuated of trapped debis by a conveyor belt, provided a bed load trap and direct quantitative measurement of bed load transport rates for comparison with bed load transport rates measured with the Helley-Smith bed load sampler.

In the vicinity of the bed load trap, the East Fork River has a bankfull discharge of about 20 m^3/s, bankfull width of 15-20 m, bankfull depth of about 1.2 m, and a slope of 0.0007. Composition of the streambed of the East Fork River at the bed load trap is primarily sand, but grav

Table 5.1. - Size distribution of composited bed material, East Fork River, Wyoming, at bed load transport research project.

Sieve diameter (mm)	Percentage, by weight, retained on sieve (%)	Percentage by weight, finer than sieve (%)
Pan	0.3	0.0
0.062	.1	.3
.088	.4	.4
.125	1.0	.8
.177	2.4	1.8
.250	6.6	4.2
.350	12.0	10.8
.500	13.5	22.8
.710	9.1	36.2
1.00	7.4	45.3
1.40	6.1	52.7
2.00	4.7	58.8
2.80	4.3	63.5
4.00	3.6	67.8
5.60	3.6	71.4
8.00	3.6	75.0
11.3	4.3	78.5
16.0	4.1	82.8
22.6	5.1	86.9
32.0	5.2	92.0
45.0	2.8	97.2
64.0	.0	100.0

bars are spaced at regular intervals of about 5-7 channel widths. Composite size-distribution data of a number of bed-material samples are listed in Table 5.1. and show ample availability of bed material for particles ranging in size from 0.25 to 32 mm.

Description of the conveyor-belt bed load trap. Across the East Fork River, a concrete trough was constructed in the bed, orthogonal to the flow direction, that would constitute an open slot, 0.25 m wide, into which any sediment moving near or on the streambed would fall. The trough is 0.4 m wide and

0.6 m deep; the level of the lip or top surface corresponds
to the natural bed, lower in elevation at the thalweg than
near the banks.

Along the bottom of the concrete trough passes an
enless rubber belt, 0.3 m wide; it is threaded around
drive and guidance cylinders, then returns overhead, where
it is supported by a suspension bridge across the river.
Thus, sediment falling into the open slot drops onto the
moving belt, then is carried laterally to a sump constructed
in the riverbank, where it is scraped off the belt. From
the sump, sediment is excavated by a series of perforated
buckets on an endless belt. The buckets lift the sediment
to an elevation 3 m above the riverbank and dump the load
into a weighing hopper. When the hopper is periodically
evacuated by opening a bottom door, accumulated sediments
fall on a horizontal endless belt that carries it 12 m in
a downstream direction and dumps the load on a transverse
endless belt; this belt, in turn, carries the debris
toward the river and dumps it into the flowing water, to
be carried downstream in a normal manner. In this way,
trapped sediment is collected, weighed continuously, and
returned to the river.

Samples of the trapped sediment for size analysis are
scooped from the endless belt as the weighing hopper is
periodically emptied. For comparison with the bed-material
size data in Table 5.1., Table 5.2. lists a transport-
weighted particle size distribution for the whole of the
bed load sampled in 1976, the most intense year of operation
The median particle size of bed load is 1.13 mm compared to
1.25 mm for bed material. Detailed data from the conveyor-
belt bed load trap have been reported by Leopold and Emmett
(1976, 1977).

Although the median particle sizes of bed load and bed
material are nearly the same, the bed material consists of
some larger particles that are rarely moved. For bed load
and bed material, Table 5.3. lists the particle size
corresponding to given particle size frequency (percentage,
by weight, finer than values). Table 5.3. clearly indicates
that some bed-material particle sizes are seldom involved in
the sediment transport process, a factor which limited the
reliability of the field tests to particle size smaller than
about 16 mm.

Sampling procedures with the Helley-Smith bed load sampler.
The spatial or cross-channel variations in bed load transport
rates are significant. Frequently, all or most of the bed
load transport occurs in a narrow part of the total width of
channel. Though this narrow width of significant transport
is generally stationary, it can shift laterally with changes
in hydraulic conditions or sediment characteristics.
Therefore, knowledge of where bed load transport has
occurred previously is not a criterion for eliminating a
portion of channel width from the sampling program. At
least 20 equally spaced, cross-channel sampling stations are
necessary to insure that zones of both maximum and minimum
transport are adequately sampled. (For large rivers and
small rivers, the technique may be modified so that sections
are not spaced greater than 15 m apart. Nor is there apparent
need for spacing sections closer than 0.5 m.).

Table 5.2. - Size distribution of transport-weighted composite bed load (1976 conveyor belt), East Fork River, Wyoming, at bed load transport research project.

Sieve diameter (mm)	Percentage, by weight, retained on sieve (%)	Percentage, by weight, finer than sieve (%)
Pan	0.3	0.0
0.062	.1	.3
.088	.2	.4
.125	.4	.6
.177	1.0	1.0
.250	5.3	1.9
.350	11.8	7.2
.500	15.1	19.0
.710	11.8	34.1
1.00	11.9	45.9
1.40	12.0	57.8
2.00	9.9	69.9
2.80	7.4	79.8
4.00	5.5	87.2
5.60	3.4	92.7
8.00	1.8	96.1
11.3	1.0	97.9
16.0	.5	98.9
22.6	.4	99.4
32.0	.2	99.8
45.0	.0	100.0

Temporal variations in bed load transport rates may also be large. This variation with time is obvious for the stream channel exhibiting movement of dunes. Even in gravel-bed rivers with no apparent dunes or migrating bedform, bed load transport may occur in slugs and show distinct cyclic trends with time.

Frequency of the cyclic trend is dependent on the velocity and wavelength of the bedform or slug of sediment. Obviously, a precise procedure would be to sample at each cross-channel station until a reliable mean transport rate was established at each cross-channel location; however, time requirements prohibit this detail.

Table 5.3. - Comparison of bed material and bed load particl
sizes.

Particle size category $(d_{(\% \text{ finer than})})$	Particle size (mm)	
	Bed material	Bed load
d_5	0.27	0.32
d_{16}	.42	.47
d_{25}	.53	.58
d_{35}	.69	.73
d_{50}	1.25	1.13
d_{65}	3.20	1.73
d_{75}	8.00	2.37
d_{84}	17.6	3.42
d_{95}	37.6	7.01

The adopted procedure (provisional method of the US
Geological Survey) is to conduct two traverses of the stream
and to sample at least 20 sections on each traverse. The
sampling duration is 30 to 60 seconds at each section. The
spatial factor is covered by the 20 sections; the temporal
factor is covered because of the time expended during a
single traverse of the stream and the time lag at each
section as the second traverse is conducted. A comparison
of values of mean transport rate, determined by multiple
traverses of the stream, shows little change in the mean
value by the addition of more than two traverses. Further,
because of changes in the river hydraulics with time, and
with each traverse of the river being time consuming, it is
often impossible to conduct more than two traverses of the
river and have the data considered as instantaneous. Each
sample collected with the Helley-Smith bed load sampler
requires about 2 to 3 minutes for lowering, sampling, raisin
emptying, and moving to a new cross-channel location. A
typical traverse thus requires about 1 hour; two traverses
require about 2 hours. The time required to complete the
double traverse generally allows a minimum of several cycles
to be sampled. In the cyclic trend of transport, this
appears adequate to average temporal variations.

In the field calibration tests, 24 cross-channel section
constituted the cross-channel frequency of sampling. Two
traverses of the stream yielded 48 individual Helley-Smith
type samples; these were averaged to give a mean bed load
transport rate, and used in the comparison with a mean bed
load transport rate for the conveyor-belt sampler.

The suspension bridge across the East Fork River at the bed load trap provided access across the river. The Helley-Smith sampler was lowered by cable to the streambed, timed for a duration of 30 seconds, and retrieved. Generally, each bed load sample was individually bagged and later air-dried, sieved, and weighed. Data thus collected could be later composited by whole-stream width for a comparison with the conveyor-belt data.

Some comparisons of results. All basic data of the field calibration tests have been summarized earlier (Emmett, 1979). Bed load transport rates were expressed as rates for each particle size class.

Relations of the bed load transport rate in each particle-size class as functions of total bed load transport rate were determined for both methods of sampling. The statistical procedure utilized was a least squares linear regression of log-transformed data, giving a power equation of the form

$$Y = AX^B$$

where Y is the bed load transport rate in a given particle size class and X is the total bed load transport rate.

Of special interest is the percentage of total bed load occurring in each particle size class and the rate of change (slope of the regression equation) in the above percentages as the actual bed load transport rate increases or decreases. These are listed in Table 5.4.

Mean percentages in Table 5.4 do not add to 100, because the mean value, $\bar{x}$, for total bedload is variable. That is, larger particles move only during higher transport rates, and the mean value of total bedload is, obviously, greater during those instances. The effect is to decrease the apparent mean percentage of total bedload in the larger particle-size classes.

Because mesh size of the sample collection bag used on the Helley-Smith sampler was 0.20 mm, data of the first two particle sized categories tabuled in Table 5.4. should be disregarded. Probably quite by coincidence, the amount of 0.06 to 0.12 mm size sediment trapped by the conveyor-belt sampler (insignificant at 0.3 percent) was nearly identical to the amount of same size material that was trapped in, rather than washed through, the Helley-Smith sample collection bag.

For sediment in the 0.25 to 0.50 mm particle size class, both samplers retain all sediment which is supplied to them. The Helley-Smith sampler showed a greater mean percentage of total bedload in this size class than did the conveyor-belt sampler. Analyses of suspended sediment data showed appreciable quantities of this size sediment in suspension. Certainly the collection of some suspended sediment by the Helley-Smith sampler is an explanation for its greater mean percentage in this size category, but a quantitative description of how much of it is attributable to this effect was not determined. It is most important to recognise that the Helley-Smith sampler does receive suspended sediment and that the absolute quantities of it are dependant on the sizes of sediment in transport and the hydraulic characteristics of the flow, facters which are different for every stream and thus cannot be calibrated.

Table 5.4   Comparison of bed load data.

Particle size class (mm)	Mean percentage of total bed load in particle size class ($\bar{Y}/\bar{X}$, in percent)		Rate of change in percentage of total bed load in particl⟨e⟩ size class (B)	
	Helley-Smith	Conveyor belt	Helley-Smith	Conveyor belt
0.06-0.12	0.35	0.32	0.727	0.663
.12- .25	3.24	1.74	.599	.553
.25- .50	22.80	18.49	.698	.742
.50-1.00	26.84	27.89	1.050	1.000
1.00-2.00	20.07	21.89	1.213	1.173
2.00-4.00	10.61	13.87	1.344	1.278
4.00-8.00	3.45	5.56	1.193	1.211
8.00-16.00	.89	1.49	.867	.995
16.00-32.00	.65	.74	.387	.926

Complete analysis of suspended sediment size data for the East Fork River showed no significant quantity of suspended sediment larger than 0.50 mm. For material capable of being moved in suspension (0.50), its significance as bed load decreases as the bed load transport rate increases. This is reflected in the rate of change values (exponent B) tabulated above. The values for suspended sediment size particles are less than unity, indicating that as total bed load transport rate increases, the percentage of sediment in those size classes decreases.

For sediment in the four particle size classes ranging in size from 0.50 to 8.0 mm, significant bed load transport occurs, and the significance increases as the total bed load transport rate increases. The dominant particle size class of bed load is 0.50 to 1.0 mm; it accounts for a little over 25% of the total bed load. The greatest rate of change in percentage of total bed load in a given particle size class occurs for particles in the size class of 2.0 to 4.0 mm

followed by size classes 1.0 to 2.0 mm and 4.0 to 8.0 mm.
These rates of change values combine with the mean
percentage values such that at high bed load transport
rates the percentage of total bed load is greatest in
particle size categories of 1.0 to 2.0 mm and 2.0 to
4.0 mm. This leads to a median particle size of composited
bed load being 1.13 mm (Table 5.3.).
    Only about 0.5 - 2% of the total bed load occurs in the
particle size categories of 8 to 16 mm and 16 to 32 mm.
The transport rate for larger particles in the East Fork
River was too minimal to allow reliable calibration for
particles larger than about 16 mm.
    The rate of change data for the two coarsest size
categories are misleading. Since the largest particles
move only at high transport rates, many low transport runs
are not included in the analysis for these size particles.
By this fact alone, large particles begin their significance
at high transport rates and increase from there. Because
zero values cannot be used in log-transformed regressions,
values of rate of change comparable to the smaller particle
size categories cannot be quantitatively determined.
    This discussion has concentrated on analysis of bed load
transport rates by individual particle size categories as
functions of total bed load transport rate. Its primary
purpose is to provide some insight into the mechanics of
bed load transport and to place reliability limits on the
comparability of data collected; it was used to show that
for particle sizes less than 0.50 mm, the influence of
suspended sediment casts doubts on comparability (not
reliability) of data collected with the Helley-Smith sampler.
For particle sizes less than 0.20 mm (mesh size of the bag),
data collected with the Helley-Smith sampler should be
discarded. For particle sizes larger than about 16 mm,
paucity of individual particles moving probably prohibits
the Helley-Smith sampler from collecting a representative
sample, and data should be treated with caution.
    Data collected concurrently with both the Helley-Smith
sampler and the conveyor-belt sampler may be compared
directly. Disregarding data for particle sizes smaller than
0.50 mm because of the suspended sediment problem, and for
particles larger than 16 mm because of the paucity of particle
problem, the comparison for each particle size class was made
with the Helley-Smith sampler results, Y, expressed as
functions of the conveyor-belt sampler results, X. As in
a previous section of this report, the statistical procedure
utilized was a least squares linear regression of log-
transformed data, giving a power equation of the form

$$Y = AX^B$$

Salient data of the analysis are given in Table 5.5.
    For particle sizes between 0.50 and 16 mm, the Helley-
Smith sampler traps approximately the same amount of sediment
as the conveyor-belt sampler. Average sampling efficiency
for those particle size classes ($\Sigma \bar{Y}/\Sigma \bar{X}$, from original
statistics, not Σ ($\bar{Y}/ \bar{W}$) from above) is 92.6%.

Table 5.5.   Comparison of concurrent bed load transport data

Particle size class (mm)	Coefficient (A)	Rate of change in ratio of transport rate (B)	Mean ratio in transport rate; Helley-Smith: Conveyor belt ($\bar{Y}/\bar{X}$, %)
0.50-1.00	0.743	0.934	98.70
1.00-2.00	.498	.868	89.36
2.00-4.00	.329	.803	86.43
4.00-8.00	.192	.739	93.81
8.00-16.00	.143	.747	93.58

If the analysis was based on values of real momentary transport rather than average values, the effect would be to increase the values of the coefficient, A, by about 8%, or to increase average sampling efficiency, $\Sigma\bar{Y}/\Sigma\bar{X}$, from 92.6% to 97.9% (cf. Emmett, 1979). A statistical correction to allow for errors in the independence variable, X, provides a correction factor of 1.07 to be applied to the exponent value, B (cf. Emmett, 1979). Finally, modifications applied to the data to allow for the operational mode of the conveyor belt (cf. Emmett, 1979) provide correction factors of 1.49 for the coefficient, A, and 1.06 for the exponent, B, giving a mean sampling efficiency, $\Sigma\bar{Y}/\Sigma\bar{X}$, of 107.7%.

Total bed load transport rates measured in the calibratio study ranged from about 0.003 to 0.3 (kg/s)/m, a range typical of many natural rivers. The bed load transport rate in each particle-size class varied from about 1 to 25% of the total rate (Table 5.4.).

For particle size classes between 0.50 mm and about 16 mm there is good agreement between the transport rate measured with the Helley-Smith sampler and that measured with the conveyor-belt sampler. Average values of the sampling efficiency range from about 90 to 110%.

Summary and recommendations

The Helley-Smith bed load sampler is a direct-measuring sampler intended for use in rivers with flow velocities up to 3 m/s and with sediment sizes up to medium gravel. Its pressure-difference design creates excessive hydraulic efficiency, about 150%; apparently this does not affect greatly the sediment-trapping efficiency of the sampler.

Sediment-trapping characteristics of the Helley-Smith bed loader sampler were studied by comparing sediment transport rates as measured with the Helley-Smith sampler with those measured utilizing an open slot constructed across a streambed. Basic data indicate that the Helley-Smith bed load sampler, for the majority of sediment sizes available in the study, is 90 to 100% efficient, but efficiency decreases somewhat with increases in transport rate. Modifications of the basic data to normalize the data sets and correct for statistical procedures indicate that the trap efficiency is 100 to 110% and varies little with changes in transport rate.

The following recommendations are made relative to the sediment-trapping characteristics of the Helley-Smith bed load sampler:

1. The Helley-Smith bed load sampler should not be used for sediment particles smaller than 0.25 mm;

2. The Helley-Smith bed load sampler should not be used for measuring bed load transport rates for sediment of particle sizes which also are transported as suspended sediment; this generally restricts use to particle sizes larger than 0.50 mm;

3. The trap efficiency for sediment in the particle size class of 0.25 to 0.50 mm was indeterminate in the calibration tests; 100% is recommended for the condition of no suspended sediment transport;

4. For sediment of particle sizes larger than 0.50 mm and smaller than 16 mm, the sediment-trapping efficiency of the Helley-Smith bed load sampler may be assumed as 100% with no change in efficiency with changes in transport rate;

5. Trap efficiency for sediment particles larger than 16 mm was indeterminate in the calibration tests; reasonable sampling efficiency may be assumed for particles somewhat larger than 16 mm, but it is likely that sampling efficiency decreases as particle size approaches nozzle dimensions.

Recommended procedure for using the Helley-Smith bed load sampler requires about 20 equally spaced, cross-channel sampling locations. Each location is sampled for a duration of 30 to 60 seconds on each of two separate traverses across the river. This procedure enables determination of mean bed load transport rate, as well as providing insight into spatial and temporal variations in transport rates.

Examples of River data

The Helley-Smith bed load sampler has been used to measure bed load in a variety of rivers ranging in channel size from less than 4 m wide to more than 6000 m wide, and transporting bed load ranging in size from medium sand to coarse gravel.

Measured transport rates have ranged from zero to about
0.5 (kg/s)/m. The most complete data set for streams
transporting primarily sand-size bed load is the information
collected for the calibration reported in this paper.
Reference to the original paper (Emmett, 1979) or to the
discussion in the present paper provides some insight into
the bed load transport of sand. The following section
presents, in graphical form, data from some gravel bed rivers

Tanana River, Alaska

The Tanana River in the vicinity of Fairbanks, Alaska, is
a large, braided, gravel-bed river. At about bankfull stage,
mean depth is of the order of 3 m, mean velocity is about
2 m/s, and the width is between about 400 to 600 m, the
greater widths being associated with an increased degree of
braiding. Channel slope is variable from reach to reach;
typical values are about 0.0005 m/m over less braided
reaches and about 0.001 m/m over reaches that are more
braided. The bed material is bimodal with modes at medium
sand and medium gravel. The supply of bimodal bed material
is such that at low to moderate flows, the mean particle
size of bed load reflects the medium-sand mode while at
higher flows the mean particle size of bed load generally
is in the medium-gravel range.
Sediment transport rates have been measured for the
Tanana River in the vicinity of Fairbanks since 1977. Data
for 1977 and 1978 have been published (Emmett, Burrows, and
Parks, 1978; Burrows, Parks and Emmett, 1979). These data
are plotted in Figure 5.3.
Suspended sediment data are typical for a glacially-fed
stream; that is, when water sediment are supplied from the
same source, the relation between water and sediment is well
defined. All of the suspended sediment data are in a five-
fold range surrounding the best-fit relation, between 40
and 200% of the mean or predicted value. A best-fit
relation for the bed load data is approximated as 1.5% of
the best-fit relation for suspended sediment. This
percentage is sufficiently small that bed load can also be
expressed as about 1.5% of total load. Bed load data also
are well defined; all but one data point are in a five-fold
range surrounding the best-fit relation, between 0.6 and 3%
of the best-fit relation for suspended sediment.

Snake and Clearwater Rivers, Idaho

The Snake and Clearwater Rivers in the vicinity of
Lewiston, Idaho are large, gravel-bed rivers, somewhat
confined because of canyon-like settings. At about bankfull
stage, mean depths are about 5 m, mean velocities are about
2.5 m/s, and widths about 150 to 200 m. Channel slopes are
variable with stage, but may be approximated as 0.0001 m/m
for the Snake River and 0.0005 m/m for the Clearwater River.
The bed material is bimodal with modes at medium to coarse
sand and medium to coarse gravel. As for the Tanana River,
the bimodal supply of bed material is such that for most
flows, the mean bed load particle size is in the coarse sand
range but at highest flows, the mean bed load particle size

104

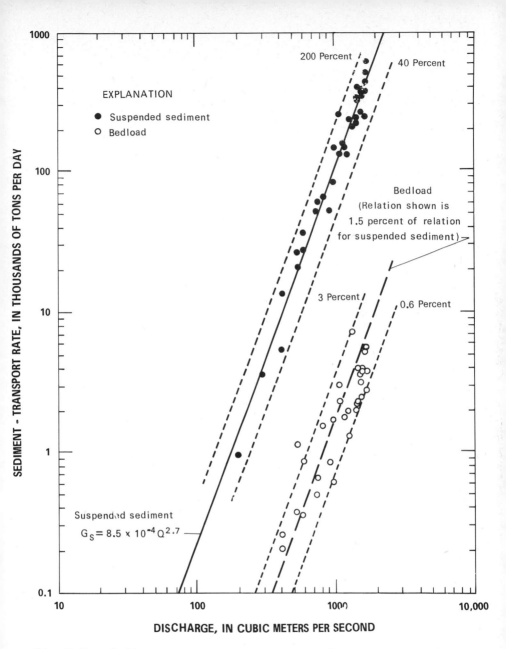

Fig 5.3. Sediment transport rate as a function of discharge, Tanana River in the vicinity of Fairbanks, Alaska.

abruptly shifts to the medium to coarse gravel range as
the streams are competent to disrupt armouring effects
and transport the coarser material.

Sediment transport rates have been measured on the
Snake and Clearwater Rivers since 1972. Data have been
published in a series of basic-data reports (Emmett and
Seitz, 1973; 1974; Seitz, 1975; 1976; Jones and Seitz,
1979). These data are plotted in Figure 5.4. Although
the two rivers have some separate attributes, their
general similarity is such that in Figure 5.4., data for
the two rivers are plotted together.

Suspended sediment data are reasonably well defined,
but are not as consistent as those for the Tanana River.
This is expected, for the wash load portions of the
suspended load come from diverse parts of the drainage area
and each tributary system has sediment transport
characteristics of its own. Somewhat fewer than two-thirds
(approximately one standard deviation) of the data are in
the five-fold range surrounding the best-fit relation.

A best-fit relation for the bed load data can be
approximated as about 5% of the best-fit relation for
suspended sediment. Indeed, the bed load data are more
consistent than the suspended sediment data; more than two-
thirds of the bed load data are in the five-fold range (2
to 10%) surrounding the 5% relation.

Discussion

Only in recent years have bed load data for rivers
existed in sufficient quantity and reliability to
facilitate a better understanding of the bed load transport
process. Data for the East Fork River, Wyoming, indicate
that when sand-size particles are dominant as bed load, the
bed load accounts for about half the total load. The
examples of the Tanana, Snake, and Clearwater Rivers
indicate that for gravel-bed rivers the bed load probably
accounts for less than 10% of the total load. But in
engineering applications of sediment data, the particle-
size coarseness of this 10% of the total load may
constitute 90% of the design problems involved in the project.

Interpretations are now being given to the bed load
data. For example, using data from the Snake and
Clearwater Rivers, Emmett (1976) has discussed the
armouring problem; Emmett and Thomas (1978) have modelled
the sediment transport process; and Bagnold (1977) has
presented ideas toward a more general understanding of
sediment transport. It may be expected that rapid advances
in sediment transport knowledge will be made as the data
base for verification of ideas is extended.

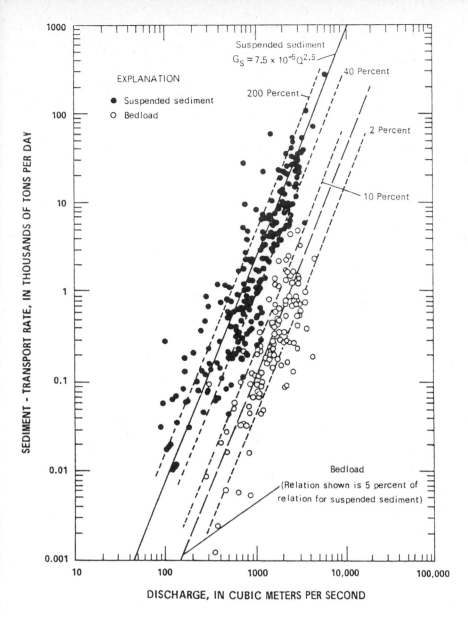

Fig 5.4.   Sediment transport rate as a function of discharge, Snake and Clearwater Rivers in the vicinity of Lewiston, Idaho.

REFERENCES

Bagnold, R.A., 1977, Bedload transport by natural rivers, *Water Resources Research*, 13, 303-312.

Burrows, R.L., Parks, B., and Emmett, W.W., 1979, Sediment transports in the Tanana River in the vicinity of Fairbanks, Alaska, 1977-78, *US Geological Survey Open-File Report* 79-1539.

Druffell, L., Emmett, W.W., Schneider, V.R., and Skinner, J.V., 1976, Laboratory hydraulic calibration of the Helley-Smith bedload sediment sampler, *US Geological Survey Open-File Report* 76-752.

Emmett, W.W., 1976, Bedload transport in two large, gravel bed rivers, Idaho and Washington, *Proceedings, Third Federal Inter-Agency Sedimentation Conference*, 4.101-114.

Emmett, W.W., 1979, A field calibration of the sediment trapping characteristics of the Helley-Smith bedload sampler, *US Geological Survey Open-File Report* 79-411.

Emmett, W.W., Burrows, R.L., and Parks, B., 1978, Sediment transport in the Tanana River in the vicinity of Fairbanks, Alaska, 1977, *US Geological Survey Open-File Report* 79-290.

Emmett, W.W., and Seitz, H.R., 1973, Suspended- and bedload sediment transport in the Snake and Clearwater Rivers in the vicinity of Lewiston, Idaho, March 1972 through June 1973, *US Geological Survey basic-data report*.

Emmett, W.W., and Seitz, H.R., 1974, Suspended- and bedload sediment transport in the Snake and Clearwater Rivers in the vicinity of Lewiston, Idaho, July 1973 through July 1984, *US Geological Survey basic-data report*.

Emmett, W.W., and Thomas, W.A., 1978, Scour and deposition in Lower Granite Reservoir, Snake and Clearwater Rivers near Lewiston, Idaho, U.S.A., *Journal of Hydraulic Research*, 16, 327-345.

Helley, E.J., and Smith, W., 1971, Development and calibration of a pressure-difference bedload sampler, *US Geological Survey open-file report*.

Hubbell, D.W., 1964, Apparatus and techniques for measuring bedload, *US Geological Survey Water-Supply Paper* 1748.

Jones, M.L., and Seitz, H.R., 1979, Suspended- and bedload-sediment transport in the Snake and Clearwater Rivers in the vicinity of Lewiston, Idaho, August 1976 through July 1978, *US Geological Survey basic-data report*.

Leopold, L.B., and Emmett, W.W., 1976, Bedload measurements, East Fork River, Wyoming, *National Academy of Science Proceedings*, 73, 1000-1004.

Leopold, L.B., and Emmett, W.W., 1977, 1976 bedload measurements, East Fork River, Wyoming, *National Academy of Science Proceedings*, 74, 2644-2648.

Schoklitsch, A., 1950 (2nd ed.), *Handbuch des Wasserbaues:* (Springer, Vienna, English translation by S.Schulits.)

Seitz, H.R., 1975, Suspended- and bedload-sediment transport in the Snake and Clearwater Rivers in the vicinity of Lewiston, Idaho, August 1974 through July 1975, *US Geological Survey basic-data report*.

Seitz, H.R., 1976, Suspended- and bedload-sediment transport in the Snake and Clearwater Rivers in the vicinity of Lewiston, Idaho, August 1975 through July 1976, *US Geological Survey basic-data report*.

6.

Dissolved loads
and their measurement

D. E. Walling

A background to dissolved loads

Although suspended sediment loads are monitored at a
considerable number of river gauging stations in countries
such as the USA, Canada, New Zealand, Roumania and the USSR,
much less attention has been given to the material
transported in solution by those rivers. In most cases it
is impossible to obtain a complementary estimate of
dissolved load. In many respects this discrepancy is
understandable, since information on suspended sediment
loadings has many important uses in the design of reservoirs
and river management schemes and in water resources
development, whilst the solute load of a river poses far
fewer problems. Nevertheless, sediment measurement
activities can in most cases be readily extended to embrace
solute loads and serious consideration should be given to
this possibility.
From the scientific standpoint, complementary information
on dissolved loads can be of great value in evaluating rates
of erosion and the relative importance of mechanical and
chemical denudation processes within a drainage basin, and
in providing a more comprehensive view of material transport
from the continents to the oceans and the global geochemical
balance. In some rivers the solution component will dominate
the total load. Measurements of the variation of solute
concentrations through the year may also afford a means of
evaluating runoff dynamics and sources and of separating the
groundwater component of the total flow (e.g. Skakalskiy,
1966; Voronkov, 1963). Furthermore, the solute load of a
drainage basin reflects catchment-wide processes and is a
sensitive indicator of catchment condition. Information on
changes in solute load through time could therefore provide
a valuable measure of human impact on the drainage basin
system and on the environment in general (Steele and Gilroy,
1971). For example, Peck (1976) has described increases in
the salinity of streams in southern Australia resulting from
the replacement of the natural forest vegetation by annual
crops and pasture; Collier *et al*. (1970) have detailed the
impact of open cast mining on solute loads; and Pierce *et al*.
(1970) have reported a classic study undertaken in the Hubbard

Reader in Geography, University of Exeter, United Kingdom.

Brook experimental catchment in New Hampshire, USA, in whi
changes in solute yield produced by forest clearance and
subsequent vegetation suppression were related to
modification of the nutrient cycle. Where irrigation is
practised measurements of dissolved load can provide a mea
of evaluating the salt balance of the irrigated area and t
resultant problems of salt accumulation and return flow
quality (e.g. Hotes and Pearson, 1977). Similarly, where
rivers discharge into lakes information on solution loads
may be of value in assessing the potential for changes in
lake productivity and for the onset of eutrophication (e.g
Environment Canada, 1976), although in this case attention
will generally be directed to the loadings of specific
nutrients and contaminants rather than to the overall
solution load.

It must also be recognised that the distinction betwee
the suspended sediment and dissolved load of a river can b
somewhat arbitrary. Colloidal material may be viewed as
extremely fine sediment (<0.45µm), but in most laboratory
procedures it will not be retained on the filter medium an
will be included in the solution phase represented by the
filtrate. In addition, interchange between the solution a
solid phases may occur during transport as a result of the
chemical mechanisms of solution, precipitation, adsorption
and desorption (e.g. Green *et al.*, 1978).

This brief overview of the nature, behaviour and
measurement of the dissolved loads of rivers will focus on
the aggregate load, often referred to as the total dissolv
solids load. The loads of individual ionic constituents w
not be considered explicitly, except in so far as a knowle
of their behaviour contributes to an understanding of the
response of the total load, although much of the material
presented will be equally applicable to them.

The nature and magnitude of dissolved loads

Table 6.1 presents information on the mean solute
composition of rivers waters of the world, based on data
contained in Livingstone (1963a). It indicates that more
than 80% of the dissolved load of rivers is generally made
up of just four components $(HCO_3^-, SO_4^{2-}, Ca^{2+}$ and $SiO_2)$,
with a number of lesser constituents comprising the remair
In some locations, however, particular constituents may as
a different importance to that shown in Table 6.1, dependi
on the nature of the catchment and the hydrological regime
(e.g. Figure 6.1A.). The material in solution may be deri
from a number of sources including precipitation and dry
deposition from the atmosphere, weathering of soil and bed
rock, mineralisation of organic matter, plant metabolism a
man-made pollution. Any attempt to relate the dissolved 1
of a river to rates of chemical denudation within the
catchment must take into account the non-denudational
component of the load. In non-limestone areas, the
bicarbonate ion (HCO_3^-) may be derived almost entirely fro
non-denudational sources associated with reactions involvi
atmospheric and soil carbon dioxide and plant metabolism.

Table 6.1. Mean major ion composition of river waters of the world (mg l^{-1}).

Region	HCO_3	SO_4	Cl	NO_3	Ca	Mg	Na	K	Fe	SiO_2	Total
N America	68	20	8	1	21	5	9	1.4	0.16	9	142
S America	31	4.8	4.9	0.7	7.2	1.5	4	2	1.4	11.9	69
Europe	95	24	6.9	3.7	31.1	5.6	5.4	1.7	0.8	7.5	182
Asia	79	8.4	8.7	0.7	18.4	5.6	9.3	-	0.01	11.7	142
Africa	43	13.5	12.1	0.8	12.5	3.8	11	-	1.3	23.2	121
Australia	31.6	2.6	10	0.05	3.9	2.7	2.9	1.4	0.3	3.9	59
World	58.4	11.2	7.8	1	15	4.1	6.3	2.3	0.67	13.1	120
World (% composition)	48.7	9.3	6.5	0.8	12.5	3.4	5.3	1.9	0.6	10.9	100

Based on Livingstone (1963a)

Zverev (1971) has estimated that the average proportion of the dissolved load of rivers in the USSR that may be attributed to precipitation and dry fallout is 14.4 per cent. It must also be recognised that dissolved ions may precipitate in some rivers in semiarid and arid environments (e.g. Blanc and Conrad, 1968) and that the dissolved load may therefore underestimate the gross output from such basins.

By examining the proportions of individual ions, particularly calcium, sodium, chloride and bicarbonate, in solution for large numbers of rivers and lakes, Gibbs (1970) proposed that three major controls influenced the chemical composition of river water. Those controls were related to the source of the solute load and were defined as firstly, atmospheric precipitation dominance, secondly, rock and soil dominance, and thirdly, evaporation - chemical precipitation dominance. According to Figure 6.1B, waters of low salt concentration (20-30 mg l^{-1} and lower) are classed as being dominated by atmospheric precipitation sources and are therefore characterised by the major cation and anion, being sodium and chloride respectively. Waters with very high salt concentrations (1000-2000 mg l^{-1} and higher) are influenced by processes of evaporation and subsequent precipitation and their composition reflects the precipitation of calcium carbonate from solution, leaving sodium and chloride as the dominant ionic constitutes. Where rock and soil sources predominate, concentrations are intermediate and the water is characterised by a calcium bicarbonate dominated

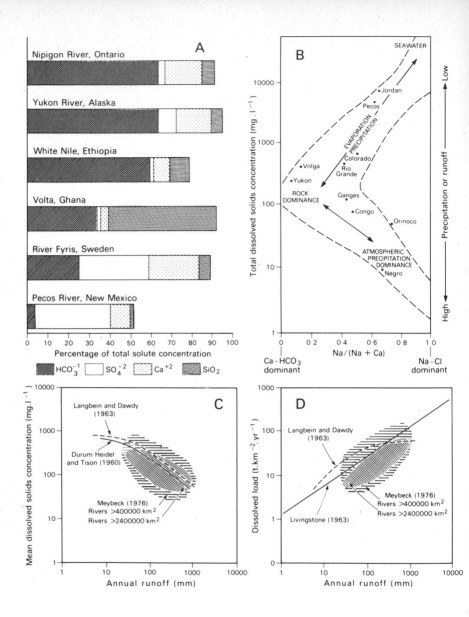

Fig 6.1. Global characteristics of river solute loads.

(A) illustrates contrasts in chemical composition (mgl⁻¹)
a number of world rivers (data from Livingstone, 1963a); (B
presents the generalised relationship between total dissol
solids concentration and ion composition proposed by Gibbs
(1970) and (C) and (D) depict relationships between dissol
solids concentration and load and annual runoff proposed b
a number of workers for U.S. and world rivers.

water chemistry. It is clear from Figure 6.1b.that there
is a large range in the proportion of these ionic
constituents for a given total dissolved solids
concentration, but this general pattern can be accepted as
a major influence on the solute composition of stream water.
Livingstone (1963a) has estimated that the mean solute
content of world river water is 120 mg l^{-1} and the related
values for individual continents are listed in Table 6.1.
Values available for individual rivers, however, range from
less than 10 mg l^{-1} for streams in the Amazon basin and in
the eastern Highlands of Victoria, Australia to 6-7000 mg
l^{-1} for rivers of Kazakhstan, USSR, described by Pavelko
and Tarasov (1967). In general, concentrations increase
with increasing aridity and this trend is clearly
demonstrated within the United States by the data presented
by Durum, Heidel and Tison (1960) for eight major drainage
areas and by Langbein and Dawdy (1964) for 170 sampling
stations, and by the world data presented by Meybeck (1976)
Figure 6.1c. Maximum solute concentrations found in arid
areas are the result of small volumes of precipitation and
runoff which become highly charged with solutes from salt
accumulations in the soil and which are further concentrated
by evaporation effects. Furthermore, the salt content of
the incident precipitation is often also at its highest
over arid areas and Matveyev and Bashmakova (1967) present
maps of the chemical content of atmospheric precipitation
over European Russia which show, for example, how along
Meridan 40°E, concentrations increase from about 10 mg l^{-1}
at 65°N to approximately 6 mg l^{-1} at 45°N. This increase is
associated with contributions from soil dust in dry areas,
an occurrence which is substantiated by a change in the
dominant chemical composition of the solute content from
sulphate and sodium ions in the north to sulphate,
bicarbonate and calcium ions in the south. Langbein and
Dawdy (1964) (Figure 6.1c.) are specific about the nature
of the decrease in solute concentrations of streams from
arid to humid conditions and argue that the decrease from
an average of 800 mg l^{-1} is gradual up to a threshold of an
annual precipitation of about 250 mm and is thereafter more
rapid, in the form of a direct dilution effect. The spatial
pattern of mean total dissolved solids concentration
exhibited by rivers in the USA and depicted in Figure 6.2
clearly demonstrates an inverse relationship between mean
annual precipitation and total solute concentration.
At a more detailed level, the average solute
concentration exhibited by a river will also reflect control
by physiographic variables such as relief and underlying
geology. Walling and Webb (1975) have described a range
of dissolved solids concentration of 25 mg l^{-1} to 650 mg l^{-1}
for small streams within the 1460 km^2 basin of the River Exe
in southwest England and the associated spatial pattern was
closely related to the geology of the area. Concentrations
of less than 100 mg l^{-1} were found in streams developed on
resistant slates and gritstones whilst values in excess of
150 mg l^{-1} and up to 700 mg l^{-1} were characteristic of less
resistant marls, sandstones and breccias. For the most
part rivers developed on igneous and metamorphic rocks
exhibit low dissolved solids concentration (< 50 mg l^{-1})

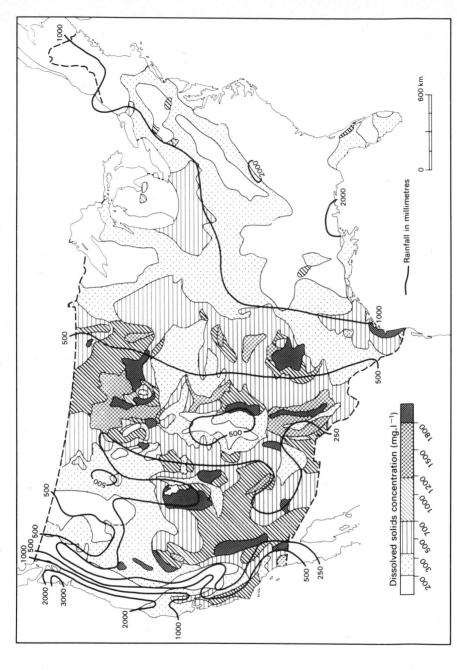

Fig 6.2. Mean total dissolved solids concentration of rivers in the United States (based on Rainwater, 1962)

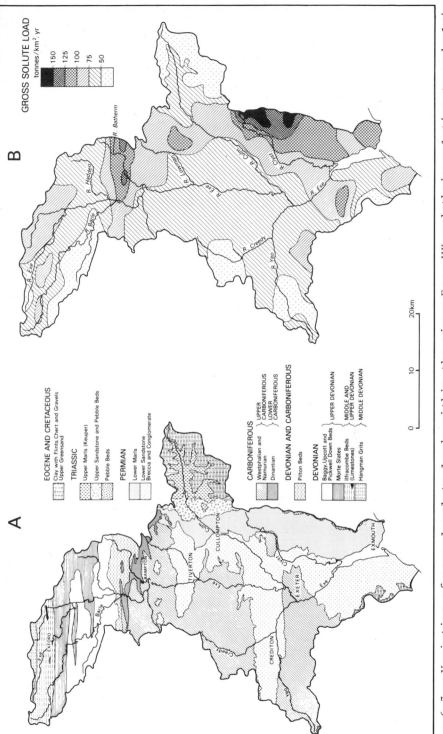

Fig 6.3. Variation of annual solute loads within the river Exe, UK and their relation to underlying geology.

whilst higher concentrations are associated with sedimentary strata. The dissolved load of a river represents the product of the dissolved solids concentration of the runoff and the volume of that runoff and will be influenced by both these variables. In general, low concentrations are balanced by high runoff volumes and global variations in dissolved load are not as great as those in total solute concentration. Strakhov (1967) reports a maximum range of 70-fold from 3.9 to 290 km^{-2} yr^{-1}. An even higher upper limit may be associated with small streams draining calcareous deposits, for Meybeck (1976) cites a solute yield of 500t km^{-2} yr^{-1} from the Dranse River in France.

The relationships between annual dissolved load and runoff suggested by Livingstone (1963b) and Meybeck (1976) for the world and by Langbein and Dawdy (1964) for the Unite States, shown in Figure 6.1d. exhibit the reverse trend to that shown for concentration (Figure 6.1c). This is becaus the decrease in concentration with increasing runoff is more than offset by the increase in runoff volume. In general, therefore, solute loads will be least in arid areas and greatest in areas of maximum precipitation. Figure 6.3b illustrates the considerable variation in solute loads that may occur within a relatively small catchment, in this case the basin of the River Exe in south west England. Variation in dissolved solids concentrations across this basin have been cited above and the pattern of solute loads similarly exhibits strong control by the underlying geology (Figure 6.3 a), with values for individual tributaries ranging from 25t km^{-2} yr^{-1} to >150 t km^{-2} yr^{-1}.

Suspended sediment and dissolved loads compared

A recent analysis of available dissolved load data for world rivers undertaken by Meybeck (1976) has produced an estimate of mean total dissolved transport from the earth's surface of 32 t km^{-2} yr^{-1}. If this is compared with global suspended sediment budgets derived by such workers as Holema (1968), it is found that there is above five times more suspended material than dissolved material delivered to the oceans. However, this ratio is not constant over the globe and the values for the individual continents derived from the work of Livingstone (1963a) and Holeman (1968) on dissolved and suspended sediment loads respectively, show considerable variation (Table 6.2.). It is clear that the value of five is strongly influenced by the extremely high sediment loads found in the rivers of Asia, since the annual dissolved load discharge from Europe exceeds that of suspend sediment. Furthermore, values of suspended sediment and tot dissolved load from individual major rivers provide many instances where the solute load exceeds the sediment load (e.g. Table 6.3.). Jaworska (1968) reports values from the Wieprz River in Poland which indicate that 95% of the total load of that river is carried in solution. However, when comparing the suspended sediment and solute loads of a river it must be recognised that the transmission losses associate with sediment transport and reflected in the tendency for sediment delivery ratios to decrease with increasing catchme

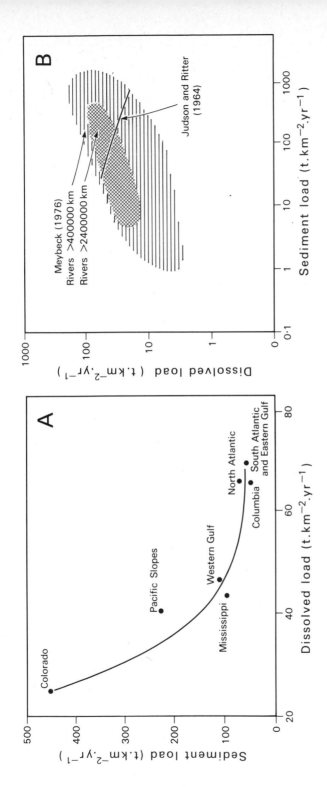

Fig 6.4. Relationships between the magnitude of the dissolved and suspended sediment components of total river load. (A) presents the relationship proposed by Judson and Ritter (1964) for US rivers and (B) depicts the relationship derived by Meybeck (1976) from an analysis of worldwide data.

119

size, are likely to be very much greater than those associated with dissolved load.

The precise relationship between the magnitude of the dissolved and suspended sediment components of total river load is one that has attracted controversy in recent years. Using data from rivers in the United States, Judson and Ritter (1964) suggested that dissolved load magnitude was inversely related to that of the suspended sediment load. (Figure 6.4a)

Table 6.2. A comparison of suspended sediment and dissolve solids discharge from the continents.

Continent	Suspended sediment discharge [1]	Dissolved load discharge[2]	Ratio sediment/disso
North America	96 t km^{-2} yr^{-1}	33.0 t km^{-2} yr^{-1}	2.9
Europe	35	42.6	0.8
Asia	600	32.2	18.6
Africa	27	24.4	1.1
Australia	45	2.3	19.6
South America	63	28.3	2.2

[1] based on Holeman (1968)

[2] based on Livingstone (1963a)

However, Soviet workers including Strakhov (1967) and Alekin and Brazhnikova (1960) believe that solute and sediment load are positively related. The more comprehensive data base available to Meybeck (1976) for world rivers has afforded a means of resolving the conflict since this also demonstrate a positive relationship between the two load components (Figure 6.4b.). It would appear that Judson and Ritter's results are something of an exception when viewed in the context of world variation which covers a greater range of load values.

Meybeck (1976) has interpreted the trend evidenced in Figure 6.4b. in terms of the influence of relief, both transport rates being at a maximum for high relief basins and at a minimum for areas of low relief. The effect of climate is superimposed on the influence of relief and is largely responsible for the scatter in the data plot. If relief is held constant, the two load components will tend to be invereseley related, since dissolved loads tend to be positively related to annual runoff (Figure 6.1d.) and sediment loads are often inversely related (c.f. Langbein and Schumm, 1958). By using five major drainage basins to derive their relationship, Judson and Ritter were dealing primarily with the influence of climate and the tendency

towards an inverse relationship between two load components
is therefore not unexpected. Meybeck (1976) has attempted
to synthesise the effects of both relief and climate on the
relative magnitude of dissolved and suspended sediment loads
and the result is shown in Table 6.4.

e 6.3.   A comparison of suspended sediment and total dissolved
loads for a number of major rivers

River	Area $(x10^3 \ km^2)$	Suspended sediment yield $(t \ km^{-2} \ yr^{-1})$	Total dissolved load $(t \ km^{-2} \ yr^{-1})$	Dissolved load as % total load
˙ado	635	870	23	2.6
ˌe	1000	150	12	7.4
ˌaputra	580	1370	130	8.7
ˌ	950	500	65	11.5
ˌs	975	537	78	12.7
ˌzi	1340	75	11.5	13.3
ˌg	795	435	75	14.7
ˌn	6300	79	46.4	37.0
ˌ	4000	13.2	11.7	47.0
ˌbia	670	43	52	54.7
	420	13.6	33	70.8
ˌ	1350	19	57	75.0
ˌurence	1025	5.0	53	91.3

ˌ on Meybeck (1976)

Characteristics of dissolved load response

   Whereas suspended sediment concentrations tend to
increase markedly with increasing discharge at a measuring
site, the total dissolved solids concentration will in nearly
all cases decline and exhibit a much smaller range of
variation (e.g. Figure 6.5a.). This inverse relation to
discharge is generally accounted for in terms of the
dilution of solute rich baseflow by water with shorter
residence times within the basin and therefore lower

121

Table 6.4. Dissolved solids and suspended sediment transport by major rivers according to morphoclimatic controls

	Dissolved Transport (t km^{-2} yr^{-1})	Sediment Transport (t km^{-2} yr^{-1})	Ratio Sediment: dissolved	Mean Annual Runoff (1 s^{-1} km^{-2})	Examples
Mountainous area, high precipitation	100-500	200-1500	2-10	20-40	Brahmaputra, Ucayali, Magdalena, Rhone
Mountainous area, low precipitation	10-80	100-1000	5-30	1-5	Colorado, Amu-Daria
Average relief, temperate or tropical climate	40-100	40-200	1-5	5-25	Danube, Mississippi, Parana Madeira
Low relief, dry climate	3-10	10-100	2-10	0.2-2	Chari, Murray
Low relief, temperate climate	20-80	20-50	0.1-1	5-15	Rhine, Northern Dvina, Volga
Low relief, subarctic Climate	10-40	1.5-15	0.1-1	5-10	Lena, St Lawrence, Finland
Low relief, tropical	3-20	1-10	0.1-1	14-60	Congo, Negro, Tapajos

Source: Meybeck (1976)

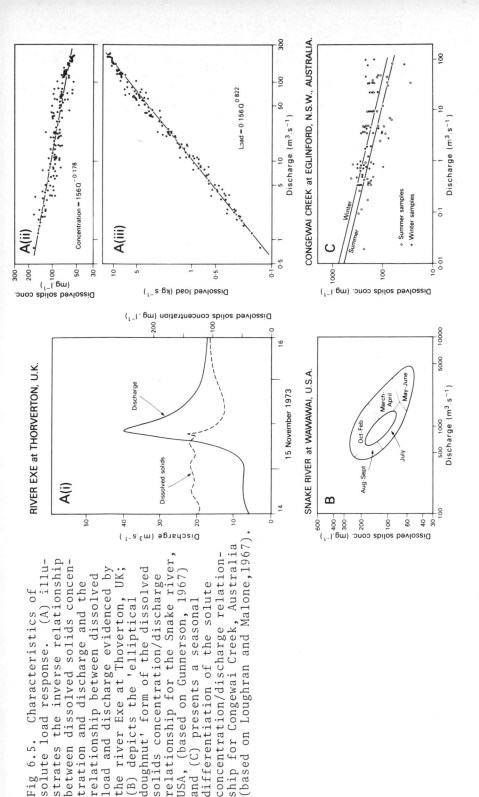

Fig 6.5. Characteristics of solute load response. (A) illustrates the inverse relationship between dissolved solids concentration and discharge and the relationship between dissolved load and discharge evidenced by the river Exe at Thoverton, UK; (B) depicts the 'elliptical doughnut' form of the dissolved solids concentration/discharge relationship for the Snake river, USA, (based on Gunnerson, 1967) and (C) presents a seasonal differentiation of the solute concentration/discharge relationship for Congewai Creek, Australia (based on Loughran and Malone, 1967).

concentrations of dissolved material. The dependence of concentration on dicharge has prompted a number of workers to develop numerical relationships between the two variables as a means of representing this prossess base. In some studies, (e.g. Duram, 1953) a hyperbolic relationsh between concentration (C) and discharge (Q) has been used, viz:-

$$C = \frac{K}{Q}$$

but in most rivers a logarithmic or simple power function o the form

$$\log C = \log a - b \log Q$$

or

$$C = aQ^{-b}$$

is applicable (Figure 6.5a.). In some instances relationsh of this type will be associated with correlation coefficien as high as 0.95 (e.g. Steele 1969), but in most cases the degree of statistical explanation will be lower. Similar logarithmic functions have also been widely applied to the relationship between instantaneous solute load (L) and discharge (Figure 6.5a.). In this case the exponent in the equation $L = aQ^b$ is positive since the decrease in concentration with increasing discharge is more than offset by the increase in water volume. The exponent in the load is by definition

$$b + 1$$

where b = the exponent in the concentration/discharge plot for the same data set.

The poor degree of explanation provided by the simple solut concentration/discharge relationship at many sites has encouraged the use of more complex functions such as polynomials and mixing model equations (e.g. Hall, 1970, 1971) and a search for equations incorporating additional variables reflecting the behaviour of the hydrological syst involved. Working on 10 years of data from the Canadian River at Whitefield, Oklahoma, Ledbetter and Gloyna (1964) proposed the relationship

$$C = a Q^b$$

where b is not a constant but is defined as

$$b = f + g \log Aq + hQ^n$$

where: Q = water discharge

a, f, g, h, n, = constants and exponents

Aq = an antecedent flow index calculated as

$$\sum_{i=1}^{30} \frac{Qi}{i}$$

where: i is the number of days back from the present.

This equation takes antecedent flow conditions into account and another approach has been to relate the solute

concentration to the relative magnitude of the various flow
components comprising the total discharge. Thus Hart *et al.*
(1964) proposed a relationship of the form

$$\text{Load} = a_1 Q_g^{b_1} + a_2 Q_i^{b_2} + a_3 Q_s^{b_3}$$

where: Q_g = the groundwater component of total flow

Q_i = the interflow component

Q_s = the surface flow component

a, b = regression constant and coefficient.

It is clearly difficult to estimate the magnitude of the
various runoff components contributing to the total streamflow
and these workers used a computer-based hydrological
simulation model, to generate the required values of the three
components. Similar in nature is the equation proposed by
Pionke *et al.* (1972) for emphemeral streams in the southwest
USA which took the form

$$\log C = a - b \log Q_b - c \, Q_s/Q_b$$

where: Q_b = base flow

Q_s = surface flow

a, b, c = regression constant and coefficients.

The use of simple mass balance models (e.g. Hem, 1970) to
account for the variation of solute concentration with
discharge exemplifies another less empirical approach to the
inclusion of flow component values in a predictive equation.
Archer *et al.* (1968) working in the Erie-Niagara basin, USA,
used a mass balance mixing model of the form

$$C = \frac{Q_g \, C_g + Q_o \, C_o}{Q}$$

where: Q_g = groundwater flow

C_g = solute concentration in groundwater flow

Q_o = overland flow

C_o = solute concentration in overland flow

In this work, the overland and groundwater flows were
estimated from analysis of the streamflow hydrograph and of
groundwater levels, and the concentration values, which were
treated as constant, were estimated from measurements taken
at times of extreme high and low flow. Another mixing model
developed by Johnson *et al.* (1969) in a study of the Hubbard
Brook catchment in New Hampshire, USA, provides an example
of the application of a mixing model to catchment stores
rather than to channel flow. They attempted to explain
variations in the solute content of streamflow as a mixing
of rainwater with soil water. Assuming that the volume
of water added to the system was directly proportional to
the flow (Q) they arrived at an equation of the form

$$C = \frac{C_s - C_r}{1 + AQ} + C_r$$

where: C_s = a constant soil water concentration

C_r = a constant rain water concentration

A = a constant inversely related to the groundwater storage volume.

In some cases, the trend of decreasing concentration with increasing water discharge may be complicated by a seasonal effect, with deviations from a simple logarithmic plot related to time of year. Gunnerson (1967) reports data from the Columbia River, USA, which exhibited this characteristic and he found that the resultant pattern of deviations exhibited an 'elliptical doughnut' form (Figure 6.5b.), indicating that the concentration/discharge relationship is dependent on the time of year. The pattern exhibited by Figure 6.5b. was explained in terms of the washing out of accumulated salts during the early part of the rainy season (late Winter to early Spring), the exhaustion of this supply and consequent dilution of concentration during late Spring and early Summer and a recovery during late Summer. Other workers such as Froehlic (1975) have found it possible to isolate seasonal variation in the concentration/discharge relationship and also contrasts between periods of snow-melt and more normal rainfall floods. Figure 6.5c. illustrates the distinction between Summer and Winter rating plots for Congewai Creek in New South Wales, Australia, as established by Loughran and Malone (1976).

Similar deviations from a simple dilution effect or straight line logarithmic concentration/discharge plot may be evidenced by the variation of dissolved solids concentrations during individual flood events. Figure 6.6a illustrates a typical situation of this nature described by Hendrickson and Krieger (1964) for streams in the Blue Grass Region of Kentucky, USA. The clockwise loop or hysteresis effect results from concentrations tending to be higher on the rising limb than on the falling limb of the flood hydrograph. This occurrence is explained by a delay or reduction in the dilution of solute concentrations durin; the early stages of increased discharge (A - B), as a resul of the flushing of accumulated soluble material from the so. As the food continues (B - C), the decline in solute availability causes a reduction in concentrations. Concentrations increase from C - D as bank storage outflow contributes to flow and as the base flow proportion increas In some streams the early part of the hydrograph will be marked by a short-lived increase in dissolved solids concentration (e.g. Walling, 1974) and this occurrence has frequently been referred to as a flushing effect.

A reverse situation to the clockwise hysteretic loop described above has been reported by Toler (1965) working o Spring Creek in southwest Georgia, USA (Figure 6.6b.).

126

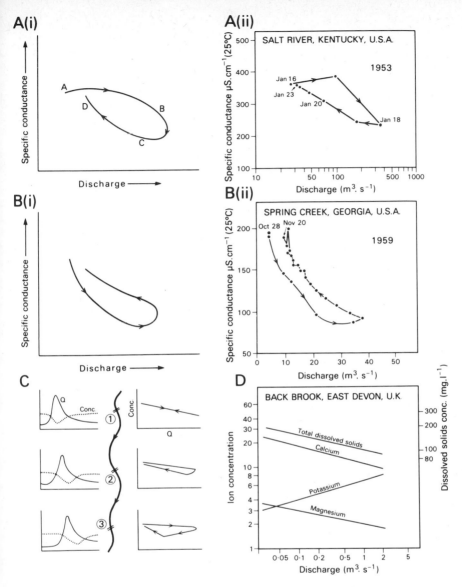

Fig 6.6. Further features of the solute concentration/discharge relationship. (A) illustrates the clockwise hysteretic form of the relationship
for the Salt River, Kentucky, (based on Hendrickson and Krieger, 1964),
whilst (B) depicts the anti-clockwise hysteretic form reported for Spring
Creek, Georgia (based on Toler, 1965). (C) indicates the effect of a
progressive lag of the minimum solute concentration after the flood peak,
produced by channel routing, on the solute concentration/discharge relationship and (D) illustrates how the response of individual ions may contrast
with that of the total solute concentration.

The cyclic relationship is still evident, but on this river concentrations
were higher during the falling stage than during the rising stage, giving
an anticlockwise hysteretic loop. This effect was attributed to a greater
proportion of groundwater discharge into the stream during the period of
falling stage which was in turn related to rapidly responding groundwater
levels associated with a good hydraulic connection between the groundwater
reservoir and the stream.

Hysteresis effects during individual floods may also be related to a lack of correspondence between peak water discharge and minimum solute concentration related to the process of flood routing (e.g. Glover and Johnson, 1974). Thus the flood wave itself may move faster than the body of water by which it was initially generated and under these circumstances the minimum concentration will progressively lag the discharge peak adding hysteresis to the concentration/discharge relationship (Figure 6.6c.).

Although it is convenient to discuss the characteristic of dissolved load response in terms of total dissolved solids concentrations it must be recognized that this measure is a lumped parameter that reflects the summation of the individual ionic constituents. These constituents may resond differently to the controls of changing discharg and season and may therefore produce an apparent complexity in the aggregate solute response of a stream. Not all ions will exhibit an inverse relationship with discharge, and chloride,nitrate, phosphate, potassium and sulphate concentrations have frequently been found to increase with increasing flow. Figure 6.6d. demonstrates how calcium and magnesium concentrations within a small stream in East Devo UK, follow a similar dilution trend to that of total dissolved solids, whereas potassium levels are positively related to discharge. Certain ions maintain near constant levels of concentration over the range of flows and this occurrence may be the result of chemical buffering mechanis within the system. Johnson *et al.* (1969) studied the variations in major ion concentrations in the stream water of the Hubbard Brook Experimental Watershed in New Hampshir USA, and distinguished three major groups of ions; those th diluted with increasing stream discharge, those that concentrated, and those that exhibited limited extremes and we possibly subject to chemical buffering mechanisms (Table 6. These contrasts in response may be attributed to the differ sources of the individual ions and will vary from catchment catchment. It is not unexpected, therefore, that individua ions will in many cases exhibit a greater degree of variati than the aggregate total solute concentration.

Similarly, ion ratios (the ratio of the concentration c an individual ionic constituent to that of the total dissol solids) must not be viewed as remaining constant under different flow conditions and at different times of the yea in response to a simple dilution effect. Lane (1975) has calculated both cation and anion ratios (defined as the rat of the ion concentration in meq l^{-1} to the total anion or cation concentration) for a number of rivers in the USA and his results indicate considerable variations between sites and between individual constituents. Figure 6.7. demonstra how the magnesium ion ratio for the Wind River in Wyoming varies very little with flow whilst that for bicarbonate in the Saline River, Kansas, ranges over nearly two orders of magnitude.

Table 6.5. General behaviour of major ions in stream water
of the Hubbard Brook Experimental Watersheds
1963-7.

Ion	Dilutes with stream discharge	Concentrates with stream discharge	Limited Extremes (chemically buffered)
Na	++	-	+
SiO_2	++	-	+
Mg	+	-	++
SO_4	+	-	++
Cl	-	-	++
Ca	+	+	+
Al	-	++	+
H	-	+	+
NO_3	-	++	+
K	-	+	+

+   irregular occurrence

++ consistent occurrence

-   not evident

Source: Johnson *et al.* (1969)

Some further comparisons with suspended sediment loads

Whereas suspended sediment concentrations tend to increase
with increasing discharge and may rise over several orders of
magnitude during flood events, dissolved solids concentrations
tend to remain relatively stable through time, but generally
exhibit a decrease during periods of increased flow (Figure
6.8A.). This contrast in behaviour has important implications
for the magnitude-frequency characteristics of the two river
load systems. Whilst a large proportion of the suspended
sediment load will be transported by the infrequent flood
flows, the transport of dissolved load will be spread more
evenly through the year. Similarly a duration curve of
suspended sediment concentration will exhibit a rather
different form to that for total dissolved solids. Figure
6.8B. presents duration curves of suspended sediment and
dissolved solids concentrations and loads for the River
Kamienica Nawojowska at Nowy Sacz in Poland contained in the
work of Froehlich (1975). Additionally, Figure 6.8C. compares
the cumulative frequency distributions of daily suspended
sediment and solute loads for one year of record from the

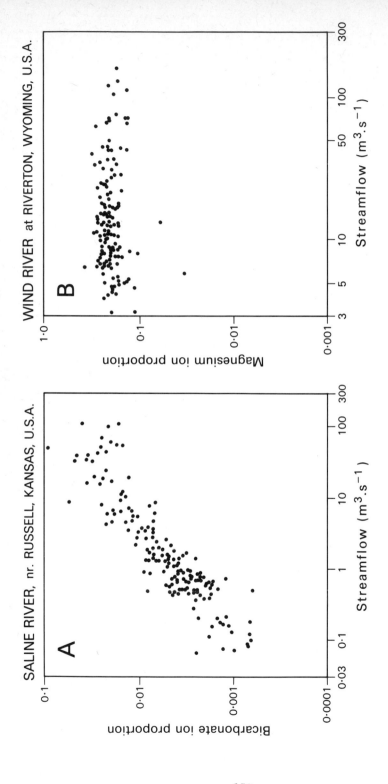

Fig 6.7. The response of bicarbonate and magnesium ion proportions in two U.S. rivers (based on Lane, 1975).

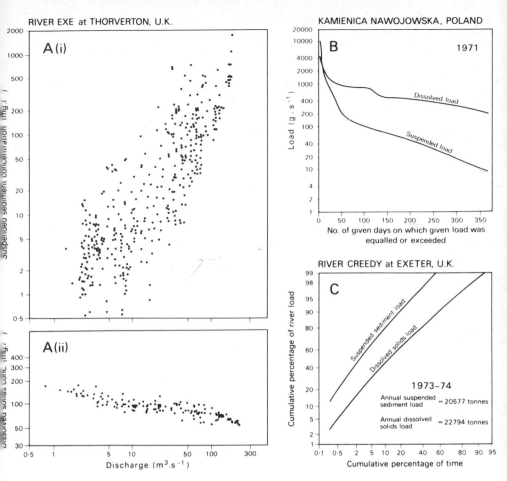

Fig 6.8. Contrasts between solute and suspended sediment response.
(A) compares the suspended sediment and dissolved solids concentration/
discharge relationships for the River Exe at Thoverton, UK, (B) contrasts
the duration curves of suspended sediment and total dissolved solids
concentration for a river in Poland (based on Froelich, 1975) and (C)
compares the cumulative frequency distribution of daily suspended
sediment and solute loads for the River Creedy, UK.

River Creedy in Devon, UK. Within this basin, nearly 50% of the annual
sediment load is transported during 2% of the time or in approximately
eight days, Whilst the equivalent figure for 50% of the annual solute
load is 48 days

The impact of man

The preceding discussion of dissolved load dynamics has been primarily concerned with an essentially natural hydrological system. It must,
however, be accepted that human activities can profoundly modify this
system. Effluent discharges from point sources may provide additional
contribution to the solute load which is unrelated to hydrological
conditions. Nonpoint or diffuse sources are more closely influenced
by the hydrological processes operating within a drainage basin, since
runoff from the land surface is generally involved in moving associated

131

pollutants to a stream. In this context one could mention
the movement of fertiliser nutrients (e.g. Omernik, 1976)
feedlot effluent (e.g. Robbins *et al.*, 1972) road salt (e.g.
Kunkle, 1972) and septic tank and sanitary landfill seepage
(e.g. Bestow, 1977) to streams by surface and subsurface
runoff.

Stormwater drainage, although often entering the stream
from a point outfall, is also best viewed as a class of
nonpoint pollution. Extremely high solute concentrations
may be associated with runoff from pavements, streets and
roads after periods of dry weather and Ellis (1975) cites
concentrations of sodium alone of 590 mg l^{-1} in a stream
receiving stormwater drainage. Receipt of stormwater
drainage highly charged with solutes by a stream during flood
events could impose a tendency for concentrations to increase
rather than decrease with increasing flow, although the
precise impact will depend on the relative magnitude of such
imputs. In contrast, a relatively constant outflow from a
point source outfall would be diluted during times of increas
flow and the stream itself would continue to exhibit the
dilution effect described in the context of natural solute
response.

Changes in precipitation chemistry, and particularly
increases in the total solute content associated with washout
and rainout from a polluted atmosphere, may also be viewed as
a source of nonpoint pollution, since the material involved
will contribute to the solute load of a stream via the basin
wide runoff processes. In addition, changes in precipitation
chemistry may themselves condition changes in the bio-
geochemical processes operating within a watershed and
controlling stream solute levels. Increased rainfall acidity
has been widely documented in northern Europe and the eastern
United States (e.g. US Dept. of Agriculture, Forest Service,
1976) and must be viewed as conditioning changes in soil and
rock weathering processes and associated solute release.

Apart from the addition of extraneous material to the
basin solute system, the impact of human activity must also
be considered in the context of modifying and accelerating
the essentially natural processes. Thus widespread disturban
of soil and regolith, such as that associated with open-cast
mining, may profoundly alter the natural weathering and
leaching processes and give rise to marked increases in
solute load. In a classic study of the Beaver Creek Basin in
Kentucky, Collier *et al.* (1970) compared a mined and an
unmined tributary and found that the rate of chemical
degradation in the mined watershed was about 12 times faster
than that in the neighbouring undisturbed basin. Similarly,
the disruption of the moisture balance occasioned by
irrigation activities, and the need to leach accumulated salt
from the soil can have a very significant influence on the
solute load of drainage water and irrigation return flow
(e.g. Hotes and Pearson, 1977). As an example, the work of
Wilcox (1962) on the Upper Rio Grande River may be cited.
This demonstrated that the weighted mean dissolved solids
concentration for the period 1934-53 increased from
221 mg l^{-1} to 1691 mg l^{-1} over a distance of 725 km as a
result of irrigation return flow. Even land use and
vegetation changes can modify the solute flux from a drainage
basin by influencing the nutrient cycle.

The classic experiment carried out in the Hubbard Brook Catchment in New Hampshire, USA, by Pierce *et al.* (1970) indicated that a somewhat severe treatment involving forest clearance and vegetation suppression could cause the total solute output from a small watershed to increase by as much as six times. Even greater increases were evidenced by certain individual ions, with nitrate loads increasing from approximately 10 kg ha^{-1} yr^{-1} to more than 450 kg ha^{-1} yr^{-1}.

The measurement of total solute concentration

The instantaneous dissolved load in a river or stream can be conveniently viewed as the product of instantaneous water discharge and the mean solute concentration in the cross section i.e.

$$L_D = \frac{Q \times C_D}{1000}$$

where L_D = Instantaneous dissolved load (kg s^{-1})

Q = Instantaneous water discharge (m^3 s^{-1})

C_D = Mean solute concentration (mg l^{-1})

Since load measurements are commonly undertaken at a streamflow gauging station, values of water discharge will generally be available and attention must be directed to the field and laboratory procedures involved in obtaining values of solute concentration. As emphasised previously, this discussion is concerned primarily with total dissolved solids concentration and does not extend to a specific treatment of laboratory analysis involving individual ionic constituents.

The collection of water samples

In the majority of rivers, turbulent mixing is sufficient to produce a homogeneous distribution of solute concentration in the cross-section. Consequently, the task of collecting a representative water sample for a laboratory analysis is generally considerably simpler than in the case of suspended sediment, where concentrations may vary markedly. Nevertheless, a number of workers have encountered variations in solute concentrations in the cross-section, particularly downstream of tributary confluences (e.g. Anderson, 1963), and surveys should be undertaken at different flow levels to ensure that homogeneity exists at the measuring site. Careful statistical analysis will be required to distinguish true cross-sectional variations from variability due to analytical errors. Hilder and Wilson (1972) describe a procedure in which samples are taken from three positions across the river, two depths are sampled at each position, each of the six samples is analysed in duplicate and the resultant data are analysed statistically using 'Analysis of Variance'. This test enables the variance due to depth and position of sampling to be distinguished from that due to random analytical error. A worked example is provided in Table 6.6.

Table 6.6. Statistical analysis of cross-section samples to
            test for homogeneity (based on Hilder and Wilson,
            1972).

A   Results

The following solute concentrations were measured

Position	Replicate Result (mg l^{-1})	
Top left	55	65
Top middle	40	45
Top right	35	30
Bottom left	45	45
Bottom middle	50	45
Bottom right	35	30

B   Analysis of Variance Table

Source of Variation	Sum of Squares	Degrees of freedom	Mean Square	F. Statistic
Depth	33.3	1	33.3	2.0
Position	816.7	2	408.4	24.4
Depth and position	216.6	2	108.3	6.5
Residual	100.0	6	16.7	-
Total	1166.7	11	-	-

C   Critical F Statistic Values

	5%	1%	0.1%
Depth	5.99	13.74	35.51
Position	5.14	10.92	27.0
Depth and Position	5.14	10.92	27.0

D   Inference

There is evidence of lack of homogeneity in the cross-
section related to position of sampling (Significant
> 99%).

If concentrations are found to be non-homogeneous in the cross-section, a sampling procedure which provides a reliable measure of flow-weighted concentration must be employed and this could make use of the techniques and instruments used for suspended sediment sampling.

Manual sampling   Where homogeneous conditions exist, a simple grab sample will provide a representative measure of solute concentration. With small streams such samples can be readily collected by wading, but in larger rivers, particularly those with steep banks, collection from a bridge may be preferable and a number of simple samplers have been developed for this purpose. Essentially, these provide a holder of sufficient weight to overcome the buoyancy of the bottle and allow it to sink (Figure 6.9). More specialised samplers such as the Meyer bottle, Dussart sampler or Kemmerer sampler (Figure 6.9.) may be used where it is desired to collect a sample at a particular depth. In the first two cases the bottle stopper is opened at the required depth by jerking the suspension cord, whilst in the latter case the sampler is closed to trap a sample representative of the chosen depth by sending a messenger weight down the suspension line. Normally, chemically neutral plastic or polythene containers should be used for sampling and sample transport. Glass may be used but it has the disadvantage of being easily breakable and there are several reports in the literature of the chemical composition of samples being altered by prolonged contact with glass. For example, Livingstone (1963a) cites a report by Hutchinson (1937) which indicated that the silica content of a water sample collected in Tibet increased from $2mg \ l^{-1}$ to $168 \ mg \ l^{-1}$ during its shipment to the USA for analysis.

Automatic sampling   In many instances, automatic samplers may be profitably employed to collect samples without the need for a visit to the measuring site. Numerous samplers of this type are now in use and these may be classified, according to the time base of collection, into interval, time-integrating, flow related, and discharge-integrating samplers and, according to the means employed to transfer the sample from the river to the storage vessel, into pump, vacuum and self-filling samplers.

Interval samplers are designed to collect discrete or gulp samples at predetermined intervals. Usually a clockwork or electric timing mechanism, which enables the operator to select the desired interval or timing, is involved. Both pump and vacuum devices are used to fill the individual sample bottles. In the former case, a peristaltic pump is commonly used to fill the sample vessel, since this avoids the problem of sample contamination by the pump mechanism. The pump is controlled by the timing device and in many instruments it is programmed to backflush the sampling line before pumping the sample. The same timing device is generally used to control the movement of a delivery spout so that it moves to the next bottle at the end of the sampling operation.

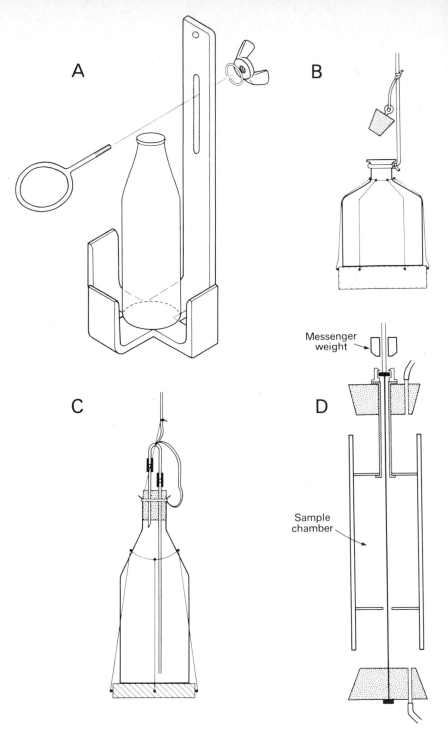

A

B

Messenger
weight

C

D

Sample
chamber

Fig 6.9.   Solute load samplers.   (A) depicts a simple depth integrating
sampler;   whilst (B), (C) and (D) respectively illustrate the Meyer
bottle, Dussart, and Kemmerer samplers which may be used to collect
samples at a particular depth.

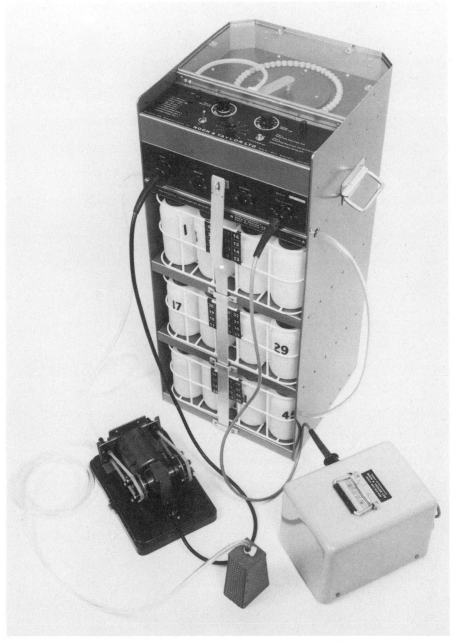

Fig 6.10. The Rock and Taylor interval sampler. Both the main sampler and the peristaltic pump are illustrated.

The Rock and Taylor interval sampler shown in Figure 6.10. is a good example of apparatus of this type. This is designed to collect 48 individual samples and provision can be made for switching the sampler through an external sensor so that it will only function when a particular threshold (eg. water level) is registered by the sensor. In this way the sampler could be preset to operate only during flood events. In some cases pump samplers of this type may deliver the discrete samples into a composite sample provided will clearly be of little value in documenting changes in solute concentration through time. With vacuum operated devices, a vacuum is created in the individual sample bottles and the timing mechanism is used to release the vacuum from each bottle in turn. Release of the vacuum causes a sample to be sucked up from the river into the bottle, since each bottle is connected to the river by an individual tube contained within a composite sampling line. Samplers of this type are limited to a suction lift of approximately 4 m and it must be recognised that peristaltic pumps will suffer from similar limitations. Figure 6.11. illustrates an interval vacuum sampler produced by Arkon Instruments Ltd., UK, to a design developed by the UK Water Pollution Research Laboratory, which is particularly suited to use in remote locations since no batteries are required.

Time-integrating samplers similarly utilise peristaltic pumps for sample collection, but in this case the flow rate is extremely low and the sample collection occupies the whole of the interval separating the filling of individual sample vessels. With a peristaltic pump, the flow rate can be readily adjusted by varying the speed of rotation of the pump or the size of the pump tube and the sample collected will integrate any fluctuation in concentration that occurred during the collection period. Simple self-filling integrating samplers have also been developed and these consist of a single collection vessel which is installed on the bed of the stream and which slowly fills through a regulated inlet valve to collect a time-integrated sample.

Flow related interval samplers can be viewed as a development of interval pump sampling apparatus in which the signal to take a discrete sample is supplied not by a clock mechanism, but by a device controlled by streamflow levels in the adjoining stream. A relatively simple system could involve the taking of a discrete sample every time the water stage changed by a given increment, whereas a more sophisticated installation might be related to increments in discharge itself. Simple rising and falling stage samplers could also be viewed as belonging to this category of apparatus. In this case, however, the sampler is self-filling and consists of a vertical suite of collecting bottles installed in the stream channel. These bottles fill automatically when the river stage exceeds (rising stage) or falls below (falling stage) the critical level associated with a particular bottle (e.g. Knedlhans, 1971).

Discharge-integrating samplers are probably the most sophisticated form of automatic sampling equipment. These are similar to flow related samplers except that the sample collection rate or timing is regulated to be directly proportional to the stream discharge. The concentration of the resulting composite sample can therefore be viewed as a discharge-weighted mean concentration for the period of collection. Regulation of the sampling rate will generally involve complex instrumentation using either a stage sensor and electronic manipulation of the stage/discharge rating function for the sampling site or a flowmeter (e.g. Claridge, 1970). Such samplers are therefore most suited for detailed research studies in small catchments and Nelson (1970)

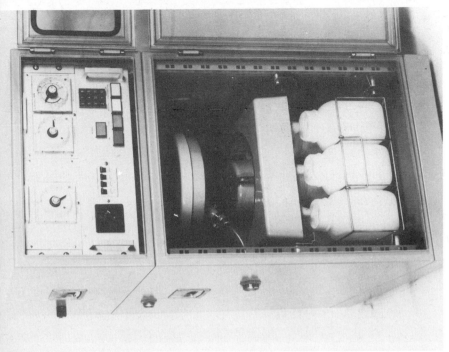

Figure 6.11. Automatic samplers. (A) illustrates the interval vacuum sampler produced by Arkon Instruments Ltd., Cheltenham, UK. and (B) shows the dicharge-integrating sampler manufactured by A. Ott of West Germany.

discusses an elaborate installation in the Walker Branch
Watershed, Tennessee which can also fractionate the sample
into up to 250 individual samples. The company A. Ott
produce a sampler of this type which incorporates several
means of sensing flow rates and which is also capable of
operating on a time interval basis (Figure 6.11.).

Continuous monitoring   Despite recent advances in sensor
technology associated with the field of continuous water
quality monitoring, there is as yet no direct means of
continuously recording total solute concentrations in a
river. However, measurements of specific electrical
conductance may be made reliably and relatively easily, and
in many cases may be used to derive a worthwhile estimate
of the continuous record of total solute concentration. Th
relationship between specific electrical conductance or
conductivity and total dissolved solids concentration will
be considered in more detail in a subsequent section.
    Modern conductivity cells, such as those utilising carb
electrodes set in an epoxy resin body, are extremely robust
and stable and may be used either as part of a pumped syste
where water is circulated from the river through a measurin
chamber, or mounted directly in a river. In the latter cas
provision must be made for retrieving the cell for cleaning
and calibration, although cell fouling is relatively
unimportant where recently developed multi-electrode cell
systems are used. Specific conductance is temperature
dependent and a water temperature sensor must be incorporat
either within the cell or close to it, so that the records
may be automatically corrected to a reference temperature o
the temperature recorded to permit subsequent manual
correction. A simple conductivity recording system involvi
a cell for mounting directly in the river, a temperature
sensor, a monitoring unit which provides automatic temperat
compensation and a recorder is shown in Figure 6.12.
Battery operated systems may also be employed and the
analogue recorder could readily be replaced by a data logge
In many cases records of specific conductance may be obtain
from exisiting water quality monitoring equipment (Figure
6.13.).
    Automatic samplers and the continuous recording of
specific conductance can vastly extend the level of samplir
which is feasible at a particular site. However, operatior
failures must be borne in mind and equipment should be
visited on a regular and frequent basis to check for
malfunctioning. Furthermore, the problems associated with
cross-sectional homogeneity discussed in the context of
manual sampling apply equally to automatic sampling and
monitoring. In addition, regular comparisons between in-
stream solute concentrations and those in the sample
container should be made to ensure that samples are
representative of ambient concentrations and are not being
contaminated by the sampler or the intake line.

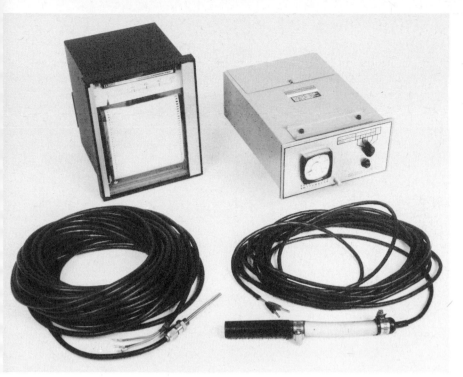

Fig 6.12. Simple conductivity recording system. In this
case the conductivity cell and temperature sensor are mounted
directly in the river and linked by cable to the monitoring
unit and strip chart recorder.

Laboratory analysis of samples

     Samples should be analysed as soon as possible after
arrival in the laboratory to minimize changes in total
solute concentration associated with precipitation and
dissolution and organic growth during storage. Ideally the
delay in analysis should not exceed one or two days but it
must be accepted that the infrequent collection of samples
from automatic samplers may involve considerably longer
delays. In some cases, automatic samplers incorporate
refrigerated and darkened sample storage compartments to
minimise the effects of sample deterioration which may be
quite severe for certain determinands. These problems are
not as severe for analyses of total solute concentration,
but it is adviseable to check the effects of sample storage
on the values of concentration obtained from analysis, to
highlight potential problems. Filtering of samples to
remove suspended solids is necessary before analysis is
undertaken and since the distinction between suspended and
dissolved solids conventionally involves a threshold of
0.45-μm, a 45 μm membrane filter should ideally be used for
this purpose. Subsequent analysis of total dissolved solids
concentration traditionally involves a residue on evaporation

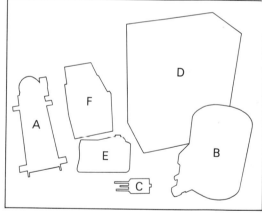

A = Pump to circulate water

B = Flow chamber with sensors

C = pH sensor

D = Monitor consol

E = Punch tape recorder

F = Strip chart recorder

Fig 6.13. An automatic water quality monitor. This system is of the type employed by the US geological survey and comprises a pump for circulating water from the river through the flow chamber, a flow chamb sensors, monitor console and a strip chart or punched tape recording de

method but the alternatives involving summation of individual constitue and estimation based on measurements of specific electrical conductance will also be briefly considered.

Residue on Evaporation Methods The total concentration of dissolved material in a filtered water sample is frequently determined from the weight of dry residue remaining after evaporation to dryness of a nowr volume of the sample (i.e. mg l^{-1}). In principal the method involved is simple, but several problems exist in obtaining precise results and interpretating the values obtained. Operational problems include the

need for accurate weighing of the empty evaporating
vessel and of the final weight of the vessel plus residue,
the need to ensure that the residue has been completely
dried and that a pocket of moisture is not contained beneath
the crust, and the need for careful use of a desiccator to
prevent the dry residue absorbing moisture during cooling.
A platinum evaporating dish is recommended (Brown *et al*.,
1970) to avoid errors associated with change in weight of
the dish and both steam baths and evaporating ovens operated
at 105°C have been used for the determination. After
evaporation, the sample residue should be further dried for
a period of 1-2 hours and whereas certain authorities
recommend final drying at 103-105°C (American Public Health
Association, 1965), it is equally common to find a temperature
of 180°C specified (e.g. Rainwater and Thatcher, 1960). The
temperature used should be clearly stated when reporting the
result.

When interpreting the result in terms of the original
composition of the sample, two points must be taken into
account. First, bicarbonate ions are unstable at 100°C and
will be converted to carbonate ions and carbon dioxide. The
bicarbonate ion content is therefore partly volatile and
other constituents such as organic matter and, to a lesser
extent, nitrate and boron may behave similarly. Secondly,
the final drying process will generally drive off the water
of crystallization from the residue, but the efficiency of
this process will depend upon the thickness of the residual
deposit, the ions involved and the drying temperature. Even
at 180°C certain salts such as calcium sulphate will not be
completely dehydrated and the importance of stating the
drying temperature is clear.

Summation of individual constituents   If a full chemical
analysis of the water sample is available it should prove
possible to confirm the value of total solute concentration
from an evaporation procedure. In order to undertake the
necessary summation, concentrations of all major dissolved
constituents are converted mathematically into the forms in
which they would normally exist in an anhydrous residue.
For example, bicarbonate in solution must be divided by
2.03 to give its equivalent weight as carbonate in the
residue. Differences between the values of total solute
concentration obtained by calculation and from evaporation
are quite common and may be accounted for by analytical errors,
by the presence of appreciable amounts of organic and some
inorganic materials that were not determined in the
constituent analysis, by the existence of water of hydration
in the residue, and by the loss of volatile solids during
evaporation and subsequent drying. Other factors such as the
behaviour of acid waters and iron compounds during evaporation
and lack of knowledge concerning the forms that elements will
take in the anhydrous residue add further complexity to the
relationship between the two values. An example of the lack
of correspondence that may occur between values of total
dissolved solids concentration obtained by evaporation and by
calculation is provided by data cited from the Sopchoppy River
in Florida by the United States Geological Survey. For one
sample, evaporation produced a value of 58 mg l^{-1} whilst

calculation gave a value of 11 mg l^{-1}. In this case the presence of high concentrations of dissolved organic material was probably important.

The distinction between values of total solute concentration obtained by evaporation or mathematical conversion and summation and those obtained from simple addition of reported ion concentrations must be clearly recognised. In some instances the latter values are reported as total solute concentrations in data compilation (e.g. Livingstone, 1963a and Table 6.1.) and indeed it could be argued that such figures are more meaningful in terms of river loading. Meybeck's (1976) work on dissolved load transport by major rivers utilises data based on simpl summation of constituent analyses and therefore includes th bicarbonate ion. This factor must be taken into account when, for example, his results are compared with those of other workers such as Langbein and Dawdy (1964) who used values of total dissolved solids concentration apparently obtained from an evaporation procedure.

The relationship between specific electrical conductance and total dissolved solids concentration. Measurements of specific electrical conductance or conductivity (μS cm^{-1}) reflect the ability of a water sample to conduct an electrical current, which in turn depends on the presence of a charged ionic species in solution. As ion concentratio increase, conductivity of the solution increases. Because of the ease with which measurements of specific electrical conductance may be made both in the laboratory and in the field, numerous workers have attempted to exploit this dependence to derive empirical relationships between conductivity (SC) and total dissolved solids concentration (TDS) (e.g. Hem, 1970; Walling and Webb,1975). The relationship is most commonly expressed as a linear equatio of the form:

$$TDS = K.SC$$

where K is a conversion factor which according to Hem (1970 generally assumes a value between 0.55 and 0.75. The preci value of K will vary according to the ionic composition of the sample, because individual ions exhibit different ion conductances (Table 6.7.). For this reason the relationshi will vary according to the sampling location, and although might also be expected to vary according to the discharge a the time of sampling, in response to changes in ion proport (cf. Figure 6.7.), many rivers exhibit a consistent relationship over the range of flows experienced (Figure 6.1

Attempts to verify empirically-derived relationships between total solute concentration and specific electrical conductance based on theoretical calculations involving the individual constituents and their ion conductances often produce poor results. Reasons for this lack of corresponde include lack of complete dissociation of the constituents (presence of ion pairs and complexes), the influence of hydr and hydroxyl ion activity on conductivity in highly acid or alkaline waters, and the presence of organic material, whic may include negatively charged poly-electrolytes, and of no

144

Table 6.7. Approximate ion conductance values for a number of common ions.

Ion	Equivalent conductivity for a concentration of 1 mg l^{-1} at $25^{\circ}C$ (μS cm^{-1})
Na^+	2.21
K^+	1.88
NH_4^+	4.08
Mg^{2+}	4.35
Ca^{2+}	2.98
Cl^-	2.15
NO_3^-	1.15
HCO_3^-	0.73
SO_4^{2-}	1.67

Based on Harned and Owen (1958)

ionized silica. In many instances silica concentrations may be appreciable and a decrease in ionic conductance values with an increase in the ionic concentration is well documented. Where TDS/specific conductance relationships do not exhibit a zero intercept, this feature may be interpreted in terms of the above factors. Thus a positive intercept could be related to the presence of un-dissociated ions or of a significant silica content and a negetive intercept could result from the presence of volatile material lost in the evaporation process.

Because specific electrical conductance values are temperature dependent and increase with increasing temperature, they are usually corrected to a standard reference temperature for reporting. A temperature of $25^{\circ}C$ is commonly used for this purpose. Ideally, samples should be stabilised at this temperature before conductivity measurements are taken, but an alternative procedure, and one that must be employed with continuous monitoring, is to apply a temperature correction factor based on the temperature of the sample. For most waters this will correspond to a temperature coefficient of 2 to 2.5% per $^{\circ}C$. The precise value may be estimated empirically, but a value of 2% is often applied arbitrarily and should not introduce large errors. Using the 2% value as an example, the necessary correction may be achieved by using one of the accepted formulae listed, viz:

145

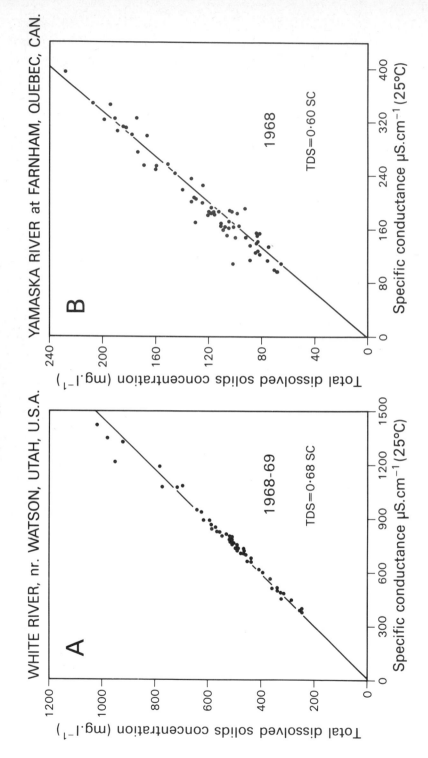

Fig 6.14. Relationships between total dissolved solids concentration and specific electrical conductance (based on data contained in US Geological Survey Water Supply Papers (A), and in Annuaires Qualités des Eaux (B)).

$$SC_{25} = SC_t \times 1.02^{(25-t)}$$

$$SC_{25} = SC_t \ (1 - 0.02(t-25))$$

where SC_{25} = Specific conductance at $25°C$ ($\mu S \ cm^{-1}$)

SC_t = Specific conductance measured at temperature
t ($\mu S \ cm^{-1}$)

t = Temperature of sample ($°C$)

It must, however, be appreciated that these two formulae produce rather different results if the temperature deviates markedly from $25°C$

Determination of the organic component of total solute concentration   In certain studies it may be of value to apportion the total dissolved solids concentration obtained by evaporation between organic and inorganic constituents. Information of this form can be used to evaluate the major sources of dissolved material in a river. A number of methods could be employed for this purpose and several of these may be briefly reviewed. No simple quantitative method for the determination of dissolved organic matter concentrations exists, but if an automated DOC (dissolved organic carbon) analyser is available this will provide a relatively easy means of determining the concentration of dissolved organic carbon in a water sample. This value may in turn be converted to a concentration of dissolved organic matter by using a conversion factor based on the carbon content of the constituent organic compounds. Values of 1.7 to 2.5 are to be found in the literature, and, in the absence of precise knowledge of the forms of organic material involved, a value of 2.0 is frequently used.

Measurements of chemical oxygen demand (COD) may also be used as a basis for estimating organic matter concentrations. Again automatic COD analysers are available, but manual methods involving wet-oxidation with potassium dichromate may also be employed (e.g. Maciolek, 1962; Menzel and Vaccaro, 1964). In this case the oxidation is generally undertaken on the residue of the sample after evaporation and this process should be undertaken at low temperatures to avoid loss of volatile organic matter. The COD value must again be converted to an equivalent value of organic matter concentration and a conversion factor of 0.7 is often used (Maciolek, 1972, Arnett, 1978). Other methods of estimating DOC, and therefore concentrations of dissolved organic material in water samples, include the derivation of empirical relationships with the correlated properties of absorbance of ultraviolet light (e.g. Dobbs *et al.*, 1972; Mattson *et al.*, 1974) and fluorescence (e.g. Smart *et. al.*, 1976.

Table 6.8. Average DOC concentrations for a number of rivers
in the USA (1969-70)

River	DOC Cocentration (mg l^{-1})
Brazos	3.3
Mississippi	3.4
Missouri	4.6
Neuse	7.1
Ohio	3.1
Sopchoppy	27

Source: Malcolm and Durum (1976)

Values of DOC concentration reported by Malcolm and Durum
(1976) for a number of rivers in the United States (Table 6.8
indicate that concentrations of dissolved organic material ma
be relatively low, but nevertheless significant, in most
rivers. This conclusion is further supported by the work of
Arnett (1978) on rivers in Yorkshire, UK, where mean dissolve
organic solids concentrations varied between 8.4 and 18.9 mg
l^{-1} and constituted between 2.7 and 10.4% of the total
dissolved load: Brinson (1976) cites similar concentrations
for tropical streams in Guatamala, although it would seem
that even higher levels may sometimes be evident, for he
reports a dissolved organic solids concentration of 36.4 mg
l^{-1} for the Oscuro river.

The assessment of solute loads

Approaches and problems

Instantaneous solute loads can be readily calculated as
the product of sample concentration and the instantaneous
discharge at the time of sampling. If data are available,
longer term loads may again be readily evaluated by combining
a continuous streamflow record with the continuous record of
total dissolved solids concentration derived from recording
conductivity apparatus. For computation purposes, the
associated records are generally reduced to a series of
frequent point measurements such that no significant error
is introduced by assuming that both concentration and
discharge values remain constant either between individual
readings or between the midpoints of the intervals on either
side of a specific data point. Uniform or variable duration
intervals could be employed, and in the latter case these
would tend to reduce in duration during periods of rapidly
changing flow. The concentration record provided by a
discharge-integrating automatic sampling installation is also
particularly amenable to the calculation of longer term loads

In this case, the concentration associated with an individual sample is the discharge-weighted mean concentration for the period of collection and the load for the period may be evaluated as the product of that concentration and the total streamflow volume.

In most cases, however, continuous records of solute concentration or discharge-integrating sampling apparatus are not available and estimates of solute load must be based upon a solute record consisting of discrete values of concentration derived from manual or automatic sampling. If an intensive sampling programme has been employed to produce a detailed record of point concentration values, these may in certain cases be assumed to adequately define the continuous record and long-term loads may be evaluated using procedures applicable to continuous records. Frequent storm-period sampling will be required in these circumstances and automatic samplers can prove of great value, particularly when their operation is linked to a preset stage or discharge threshold. The required frequency will clearly be related to the catchment area of the study stream, because water discharge and related solute response may fluctuate very rapidly in a small basin (Walling, 1975), whereas the rates of change will be considerably slower in a large watershed. Where an automatic time-integrating sampler has been employed, the values of concentration associated with individual sample bottles will relate to the periods between sample collection and may serve to integrate small fluctuation in concentration.

Sampling frequency will generally be insufficient to define the continuous record of concentration reliably and load calculation must then depend on <u>interpolation</u> or <u>extrapolation</u> procedures. Because solute concentrations vary over a relatively small range, particularly when compared to suspended sediment concentration (Figure 6.8.), such procedures will not in general introduce large errors. Nevertheless care must be exercised to ensure that the calculation procedure adopted takes account of the hydrological basis of solute response. For example, estimation of annual solute load as the product of annual water discharge and the mean of a series of concentration measurements (e.g. Ongley, 1973) will in most cases overestimate the load. This is because an inverse relationship between concentration and discharge will produce a discharge - weighted mean which is lower than the simple arithmetic mean of available observations. Meybeck (1976) refers to the case of the Mekong River at Phnom Penh where the arithmetic mean total solute concentration of 121 mg 1^{-1} contrasted with a discharge-weighted mean of 95 mg 1^{-1}. Use of the former value to calculate total solute load would over-estimate the load by 27%. No standard method exists for calculating loads from a series of infrequent samples and a number of the procedures employed by various workers may be reviewed.

<u>The interval method</u>   This procedure introduces the assumption that the concentration value associated with a particular sample is representative of the ambient concentrations in the river during the period between individual samples.   More specifically the load for the period between two samples may be calculated as the product of the discharge for that period

and the mean of those two samples (the mean-interval approach), or an individual sample value may be applied to the streamflow volume associated with the time period extending half way towards the preceding and succeeding samples (the mid-interval approach). These approaches have been widely used by ecologists attempting to derive catchment nutrient budgets (e.g. Likens *et al.*, 1967) and by geomorphologists (e.g. Zeman and Slaymaker, 1978). A programme of regular weekly sampling is often employed and it must be recognised that the strategy will only provide reliable estimates of solute load when concentrations vary little in response to discharge and other temporal controls.

A variant of the interval method could be employed where automatic sampling equipment is used to collect samples at intervals proportional to flow rate, for example every time the cumulative discharge total increases by a predetermined amount, rather than at regular time intervals. The calculation procedure will be simplified in this case, since the flow volume associated with each value of concentration will remain constant. Claridge (1970) produced estimates of individual ion loads from a small New Zealand catchment, using both flow proportional and time-based (weekly) interval sampling, and found that the two estimates could differ by as much as 33% since the weekly sampling programme did not document the rapidly changing concentrations associated with periods of high discharge.

Dissolved load rating relationships   Whereas the interval method essentially involves interpolation between individual sampling occasions, load estimation techniques which make use of rating relationships involve extrapolation of the available data base. In this procedure, a relationship between total solute concentration or total solute load and discharge is developed using the sampled concentrations and associated values of instantaneous discharge and this relationship is used to estimate concentrations or loads during periods when samples were not collected. Loads may therefore be evaluated using similar calculation procedures to those employed with frequent sampling or continuous recording, because the concentration or load associated with any given time period may be estimated from the rating relationship. The form of the rating relationship will vary according to the solute response of the study stream, but, in the case of concentration straight line functions fitted to logarithmic plots of concentration versus discharge (e.g. Figure 6.5.) are frequently used. More sophisticated ratings could involve the use of polynomial (e.g. Wartiovaara, 1975) and other function the use of separate relationships for different seasons and for rising and falling stage (e.g. Loughran and Malone, 1976) and the inclusion of other variables such as rate of change of discharge in the relationship (e.g. Johnson *et al.*, 1976). some instances, a line fitted by eye to a simple graphical pl may prove more appropriate, particularly if the plot demonstrates abrupt changes in the trend of concentration versus discharge at the extremes of high and low flow. In this case, however, the lack of a mathematical function will necessitate the graphical plot itself being used for extrapolation and the rating relationship will not be readily applicable to automated data processsing.

A computer routine is frequently used to apply the rating function to hourly or half-hourly values of discharge in order to calculate the associated estimates of solute load and to total these values to produce longer term load estimates. It should be appreciated that a rating relationship based on instantaneous values of concentration and discharge is not directly applicable to estimating values of daily mean concentration from values of daily mean discharge, although this is frequently undertaken and the errors involved are likely to be small.

The computation involved in estimating loads using a rating relationship and a continuous series of short interval discharge data may be considerably reduced by using the streamflow duration curve. The method is essentially the same as that used in the flow duration - sediment rating curve technique of estimating suspended sediment loads (Miller, 1951). The discharge record is subdivided into a large number of duration classes and the load for each duration class is estimated as the product of the discharge associated with the mid-point of the duration class, the concentration estimated for that discharge value from the rating relationship, and the total duration of flows in the class (Table 6.9.). The magnitude of the duration classes should be selected with regard to the rating plot to ensure that each class represents only a small change in concentration, and the classes may be of variable duration. Again it must be recognised that many streamflow duration curves are derived from daily mean discharge data and that an inconsistency is involved if they are used with instantaneous rating relationships.

The flow-interval method of calculating river loads developed by the US Army Corps of Engineers may be viewed as a development of the rating relationship approach. In this case the graphical plot of daily load versus daily discharge is partitioned into uniform intervals or classes of flow (Figure 6.15B.) and the mean load for each interval is calculated from the plotted points. The record of daily mean flows is partitioned into the same flow intervals, and the load associated with each flow class is calculated as the product of the mean load for that flow interval and the number of days that discharges within that class occurred. The total load is obtained as the sum of the loads calculated for individual flow intervals. A further development of the method enables the standard errors of the values of mean load relating to each flow interval to be calculated and these values can in turn be used to evaluate a standard error term for the final load estimate, which takes into account the degree of scatter in the original graphical plot. Some inconsistency is introduced in the use of instantaneous samples to derive the relationship between daily load and daily flow but with most rivers this is unlikely to introduce significant errors. The choice of the number of flow intervals must again be related to the degree of variation of load according to discharge.

A combined approach   Both interpolation and extrapolation procedures may be combined in a load estimating technique if observed values of concentration or load are used to adjust the estimates obtained from a generalised rating relationship.

151

Table 6.9. Calculation of solute load using a flow duration
curve and a solute concentration/discharge rating
relationship

1	2	3	4	5	6
Duration Curve Interval (% time)	Interval Mid-point	Duration (% time)	Discharge ($m^3 s^{-1}$)	Concentration (mg l^{-1})	Load (tonnes
0-0.05	0.025	0.05	72	103	116.9
0.05-0.2	0.125	0.15	43	120	244.1
0.2-0.5	0.35	0.3	31	130	381.3
0.5-1.0	0.95	0.5	23	138	500.5
1-2	1.5	1	17.5	143	789.2
2-5	3.5	3	12.2	150	1731.3
5-10	7.5	5	7.6	156	1869.5
10-20	15	10	5.4	163	2775.8
20-30	25	10	3.7	169	1971.9
30-40	35	10	2.7	176	1498.6
40-50	45	10	2.0	183	1154.2
50-60	55	10	1.6	191	963.7
60-70	65	10	1.2	202	764.4
70-80	75	10	0.98	215	664.5
80-90	85	10	0.73	231	531.8
90-95	92.5	5	0.52	254	208.3
95-98	96.5	3	0.425	273	109.8
98-99	98.5	1	0.35	286	31.6
99-99.5	99.25	0.5	0.31	299	14.6
99.5-99.8	99.65	0.3	0.27	306	7.8
99.8-99.95	99.875	0.15	0.24	312	3.5
99.95-100	99.975	0.05	0.22	319	1.1

Mean annual load for period of record   16334.4 ton

Col. 4 = Discharge associated with interval mid-point (col. 2)

Col. 5 = Concentration derived from rating relationship for the discharge list
in col. 5

Col. 6 = Load = col. 4 x col. 5 x col.3 x 0.31536

Note. In this example a limited number of duration curve intervals has been u
a greater number would provide a more accurate result. The streamflow
duration curve and rating relationship used in this calculation are
illustrated in Figure 6.15A.

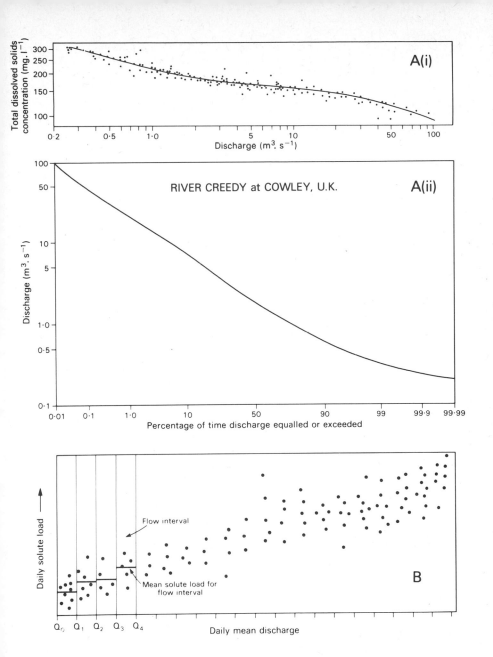

Fig 6.15. Calculation of solute loads using the flow duration-
solute rating curve and flow interval techniques. The streamflow duration and solute rating curves presented in (A) have
been used in Table 6.9. to calculate the mean annual solute
load for the River Creedy. (B) illustrates how the flow
interval technique partitions the flow record into a number
of equal intervals and how the mean solute load is determined
for each of these intervals.

Thus if samples were collected at approximately weekly intervals, the data obtained could be used to derive a rating relationship. Values associated with actual field samples could then be compared with the rating estimate for the same moment in time and be used to adjust the rating estimates for that week. If the rating relationship took the form

$$\text{Concentration (C)} = aQ^b$$

the adjusted estimate of concentration for a particular moment in time could be evaluated as

$$C = \left[\frac{C_1}{aQ_1^{\,b}} + \frac{t}{T} \times \left(\frac{C_2}{aQ_2^{\,b}} - \frac{C_1}{aQ_1^{\,b}} \right) \right] \times aQ^b$$

Where:

C Concentration to be estimated

C_1 = Concentration associated with the last field sample taken before the estimation point

C_2 = Concentration associated with the first field sample taken after the estimation time

Q_1 = Instantaneous discharge at time of sampling C

Q_2 = Instantaneous discharge at time of sampling C

T = Time interval between samples C_1 and C_2

t = Time interval between sample C_1 and the estimation point

Q = Discharge at time concentration estimate required

a,b = Constant and exponent in the rating relationship

The adjustment factor is thus interpolated between individual sampling occasions. This approach could prove particularly useful where a seasonal effect exists in the concentration versus discharge relationship for a river and where values of concentration for a given value of discharge tend to be higher at one time of the year than another. Similarly it may provide a useful adjustment where a flushing effect occurs at the end of a long dry summer.

The reliability of dissolved load estimates

Although values of dissolved load are commonly reported as absolute values, it must be recognised that they are only estimates and that they will incorporate errors. These error may be due to sample collection procedures, to analytical techniques used in the laboratory to determine solute concentration and to the methods used to evaluate loads. In the latter case errors may arise from inaccurate streamflow data, lack of frequent sampling and the shortcomings of any interpolation or extrapolation procedure involved.

Errors associated with field sampling and laboratory analysis can be quantified by replicate sampling and analysis and by inter-laboratory comparisons. Measurements of total dissolved solids concentration obtained using the residue on evaporation method may introduce errors of 10% or more because of the problems outlined in the section on laboratory analyses. Little information is, however, available on the potential errors associated with sampling frequency and load calculation procedures. Walling (1978) has attempted to investigate this topic by comparing the estimates of total solute load for two rivers in Deven, UK, obtained using various sampling strategies and calculation procedures with those obtained using continuous records of specific conductance to derive a continuous record of dissolved solids concentrations. Assuming that this latter record produced an accurate estimate of total load, errors associated with using a rating relationship devloped from a detailed sampling programme to estimate annual loads were of the order of $\pm$5% whilst estimation errors for monthly loads were generally less than $\pm$12%. The use of daily mean flow data with instantaneous rating relationship did not significantly increase errors.

Use of a regular weekly sampling programme to estimate loads using the interval method produced errors in annual load estimates within the range -2% to +8% and these were reduced to -1% to +4% when a regular daily sampling programme was evaluated. For a weekly sampling interval, the errors associated with monthly load estimates were in the range -5% to +15%, whilst for the daily programme they ranged between -1% and +4%.

These errors could be expected to be larger with less intensive sampling programmes, and further analysis of this nature is required to extend the findings to other sampling strategies and calculation procedures and to river basins with different hydrological and physiographic characteristics. Nevertheless, from these limited findings it can be suggested that estimates of annual load based on a well defined rating relationship or sampling at weekly intervals could introduce errors of $\pm$10%, whilst for monthly loads the errors could increase to $\pm$20%. These error terms should be borne in mind when the associated load data are used in statistical analyses or for evaluating solute budgets. The possibility of producing a standard error for the reported load estimate provided by the flow interval technique could prove of considerable value, even though this term does not relate to an absolute standard.

A further aspect of the reliability of solute load estimates relates to the assessment of mean annual loads and the number of years of record required to give a reliable estimate of this value. Leaving aside problems of climatic change, modifications in catchment condition, human influences on water quality and other potential causes of non-stationarity in solute response, the reliability of any estimate of long term mean load will depend upon the length of record and the natural variability of the annual load totals. A long length of record would be necessary to reach definitive conclusions as to the likely reliability of an estimate of mean annual load derived from a particular record length or as to the

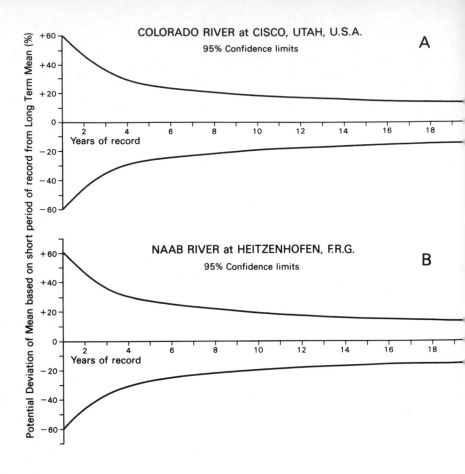

Fig 6.16. Estimates of the reliability of mean annual
solute load values based on a short period of record.

length of record required to produce an estimate within
given confidence limits. Nevertheless, some indications o
the potential errors involved are provided in Figure 6.16.
These results have been derived from analysis of relativel
short periods of record (c. 10 years) for one river in the
western USA and for one in the Federal Republic of Germany
It has been tentatively assumed that the variance of these
short periods of record provide a worthwhile estimate of t
total population variance, and standard error values relat
to particular lengths of record and confidence limits have
been calculated. The two graphs shown in Figure 6.16 are
remarkably similar, bearing in mind the contrasts between
the rivers and suggest that an estimate of mean annual
solute load based on only five years of record could lie
between ±25% of the true long term value at the 95%
confidence level.

The design of sampling programmes

Because solute concentrations generally exhibit much less variation than suspended sediment concentrations (e.g. Figure 6.8.) the problems associated with designing an effective sampling programme are much less severe. In practice, much will depend on the strategy selected for load calculation, but it is essential that the pattern of solute response exhibited by the river and its hydrological regime should be taken into account. If possible, a pilot study should be used to clarify this pattern so that a synthetic record can be generated and potential sampling strategies tested. A strategy suited to a river regime dominated by spring melt might be rather different from one developed for a river in a humid temperature zone or in an arid region. In the case of the interval method, the sampling frequency should be adequate to describe the general pattern of variation in solute concentrations through time and to prevent large changes in concentration occurring between individual samples. A variable sampling interval may prove desirable where storm flow events are concentrated into a limited portion of the year. The results discussed in the section on the reliability of dissolved load estimates for a river in the UK suggested that weekly sampling would provide worthwhile estimates of annual load for that river, although more frequent sampling would be required to produce reliable estimates of monthly load.

If the solute response of a stream is characterised by frequent and marked variations in concentration, the rating curve approach may prove more suitable for evaluating solute loads. In this case, the sampling programme must be designed to provide representative samples from the complete range of flows and contingency plans may be necessary to ensure that samples are collected during short-lived flood events. Equally the samples must provide a representative coverage of different seasons of the year to document any seasonal differentiation of response. Further, the programme must be capable of documenting any change in catchment behaviour associated with human activity and other causes of non-stationarity. When a load assessment technique is to involve both extrapolation and interpolation procedures the sampling programme designed to establish a meaningful rating relationship must be supplemented by a series of samples taken at regular intervals to provide a basis for interpolation.

The interpretation of solute loads

For certain purposes it is useful to view the solute load of a river as the product of chemical denudation and to use the load estimate to evaluate a rate of chemical denudation or solutional erosion. Use of solute load data for this purpose involves many assumptions and problems and a number of these may be briefly reviewed.

Volume and rate conversions

The conversion of solute loads (tonnes km^{-2} yr^{-1}) to equivalent volumes or depths of chemical denudation involve: values of specific gravity or density, i.e.

Denudation in m^3 km^{-2} yr^{-1} or mm $1000yr^{-1}$ = $\dfrac{\text{total load (tonnes)}}{\text{area } (km^2) \text{ x specific gravit}}$

Although Corbel (1964) has suggested that a general val of 2.5 should be employed for the specific gravity, this figure could be varied to suit individual rock types. The reporting of values in units of mm $1000yr^{-1}$ may prove misleading where chemical weathering takes place within the rock strata, soil and regolith and is compensated by a decrease in bulk density. Furthermore, the expression of denudation rates as an average for the entire watershed introduces the assumption that chemical denudation has proceeded uniformly over the catchment area. Recent acceptance of the partial-area concepts of runoff generatio: (e.g. Betson, 1964; Hewlett and Hibbert, 1967) clearly indicates that solute removal is likely to be spatially non uniform. Finlayson (1977) provides a useful discussion of this theme and suggests that denudation rates within a smal basin on the Mendip Hills, UK, varied between 1 mm $1000yr^{-1}$ and 3.5 mm $1000yr^{-1}$ in response to the pattern of runoff generation and the moisture status of the slope units.

Extrapolation of denudation rates backwards in time to obtain rates of surface lowering and of landscape developme also involves the assumption that the drainage basin system remains stationary. It is, however, clear that changes in climate have occurred in the recent past and contemporary solute loads may be far from representative of longer term conditions because of human activity (e.g. Meade, 1969).

Assessing the denudation component of the total solute load

Of crucial importance in the use of solute load data to document rates of chemical denudation is the need to determ the portion of the load which is derived from chemical weathering of soil- and rock-forming minerals. It is well known that a significant proportion of the inorganic solute load discharged from a basin may be attributed to atmospher sources, to atmosphere-plant-soil interactions, to chemical fertilizers and other non-point pollutants and to municipal and industrial effluents, and therefore cannot be ascribed the processes of chemical denudation.

Even in catchment areas with minimum human interference a sizeable proportion of the solute load will be non-denudational in origin and attributable to an atmospheric source. Material from this source will include cyclic salt from the oceans, soil dust and atmospheric pollutants and m be subtracted from the gross solute load. Meade (1969) cit studies in North Carolina and New Hampshire, USA, which indicate atmospheric contributions to dissolved loads of 20% and 50% respectively. Similarly, work by Miller (1961) in the Sangre de Cristo Mountains of New Mexico showed that th proportion of the dissolved load derived from precipitation inputs was about 50% for streams draining quartzites, 33% for streams draining granite and 10% for streams on sandstones and claystones. When the individual ionic constituents are considered, the proportions of the loads associated with bulk fallout will clearly vary. In a detailed study of the solute budget of a small forested catchment in the mountains of British Columbia, Zeman and Slaymaker (1978) attributed about 25% of the solute yield to atmospheric sources, but the values for individual ions ranged between 100% for ammonium and nitrate to 17.4% for calcium, and for silica it was as low as 0.8%.

Correction for atmospheric inputs provides an estimate of the porportion of the solute load directly related to terrestrial sources, but again a significant proportion of this load may not be derived from the weathering of mineral material (c.f. Janda, 1971). Atmosphere-soil-plant-water interactions will frequently account for the majority of the bicarbonate content of streamflow and this ion is a major constituent of most dissolved loads (Table 6.1., Figure 6.1A.). In this case carbon dioxide associated with plant metabolism, bacterial activity, and the atmosphere dissolves in rainfall and soil water to produce H_2CO_3 which in turn dissociates under normal pH conditions to give H^+ and HCO_3^- ions. Carbonic acid (H_2CO_3) is an important source of hydrogen ions in many weathering reactions and it must be recognised that the bicarbonate ions associated with the resulting soluble products are not derived from the mineral material. For example, in the dissolution of aluminium silicates which make up 70% or more of the rocks in contact with the surface and subsurface water of the hydrological cycle, the primary minerals are converted into hydrated secondary minerals with an accompanying release of cations and silicic acid e.g.

$$\text{Cation Al-silicate} + H_2CO_3 + H_2O \rightarrow HCO_3^- + H_4SiO_4^0 + \text{cation} + \text{Al-silicate}$$

Even in the case of limestone solution, half of the bicarbonate comprising the soluble product is not rock-derived, i.e.

$$CaCO_3 + H_2CO_3 \rightarrow Ca^{2+} + 2HCO_3^-$$

A large proportion of the nitrate ions contained in the terrestrial component of a solute load may similarly be associated with the soil-plant system and be derived from organic matter by the processes of mineralisation and nitrification i.e.

$$CNH_2 \text{ (organic)} \rightarrow NH_3 \qquad)$$
$$\qquad\qquad\qquad\qquad\qquad\qquad) \text{ (mineralisation)}$$
$$NH_3 + H_2O \rightarrow NH_4^+ + OH^- \quad)$$

$$NH_4^+ + 2O_2 \rightarrow NO_3^- + 2H^+ = H_2O \quad \text{(nitrification)}$$

Again the source is unrelated to weathering of mineral material.

When a drainage basin is subject to human activity, pollution from both diffuse and point sources may provide a further non-denudational component of the river load. In the case of nonpoint pollution, fertilizer nutrients and sodium and calcium chloride from road salting may, for example, be involved and a wide variety of chemical substances are associated with municipal and industrial effluent. A study of nutrient loadings in the Potomac river basin by Jaworski and Hetling (1970) has indicated that 87% of the phosphorus load was derived from point source wastewater runoff.

No generally accepted method exists for determining the rock-derived or denudational component of the total solute load. Atmospheric contributions and pollution loadings must clearly be subtracted, but detailed knowledge of soil and bedrock geochemistry may be necessary to decipher the precise importance of weathering contribution to the remaining load. The source of the bicarbonate load component must be careful evaluated since in certain cases this may be entirely non-denudational. It is possible that less than 50% of the solute load remaining after subtraction of atmospheric and pollution contributions will be rock-derived and the results of many investigations of rates of chemical weathering may need revision in the light of this fact.

Strategies for the prediction and modelling of solute loads

Activity concerned with the prediction and modelling of solute loads has been less evident than in the case of sediment yields, probably because of their lesser importance in the context of environmental and economic problems. There are no direct parallels with such problems as soil loss, reservoir sedimentation and channel management. Nevertheless the topic impinges on the broad field of water quality modelling and much of that work is of relevance. Space does not permit a lengthy and detailed discussion of proven and potential techniques for predicting and modelling solute yields, but some of the strategies involved may be reviewed. In undertaking this review, a useful distinction may be made between approaches concerned with spatial variation, those concerned with temporal variation, and those attempting to encompass both dimensions.

Prediction and modelling of spatial variations in solute yie

Gorham (1961) and subsequent writers (e.g. Douglas, 1972, Walling and Webb, 1975) have emphasised that spatial variation in solute concentrations and dissolved loads may be related control by climate, parent material, topography and biotic factors and the influence of these controls has already been touched upon. Thus the relationships between mean solute concentration and annual runoff and between dissolved load and annual runoff shown in Figures 6.1C. and D. could be viewed as simple global scale predictive models based on climate, although no indication of their reliability is provided. In the case of the relationships proposed by Meybeck (1976), the prediction would be in terms of a likely range rather than a specific value. Within this range, the magnitude of the value would depend upon the influence of the remaining three controls. The predictive capacity of simple global relationships of this type could clearly be improved by incorporating additional surrogates for climate and including variables representing the other three controls.

At a more restricted spatial scale, attempts have been made to incorporate more than one variable and the work of Biesecker and Leifeste (1975) can be cited as an example. Their work was concerned with evaluating water quality data collected from 57 hydrologic bench-mark stations distributed across the USA and included an assessment of the value of indices of climate and rock type as predictive tools for determining the maximum dissolved-solids concentration expected at a particular location under 'natural' conditions. Figure 6.17A. presents their attempt to establish relationships between mean and annual runoff and maximum dissolved solids concentration for individual rock types. This could provide the basis for a simple predictive model, although the authors emphasise that the results apply to 'natural' areas as represented by bench mark catchments and that rivers draining areas subject to human activity exhibit considerably higher solute concentrations. The impact of human activity in terms of both point and nonpoint sources of pollution could be introduced through loading functions of the type developed by McElroy *et al.* (1976) for nonpoint sources. In this work procedures for calculating the excess loads resulting from such sources as agriculture, urban runoff, construction and silviculture were developed. Thus guidelines are given for applying a simple salt balance to calculate the component of the solute load of an irrigated catchment associated with return flow as,

$$TDS_{(IRF)} = a.A.C(TDS)_{GW} \ (IRR + Pr - CU)$$

Where: $TDS_{(IRF)}$ = Solute load in irrigation return flow

A = Irrigated area

IRR = Irrigation water added to crop root annually

Pr = Annual precipitation

Cu = Annual water consumptive use

$C(TDS)_{GW}$ = Concentration of dissolved solids in groundwater contributing to subsurface return flow

a = Conversion factor

At both the countrywide and the more local level, potential exists for the inclusion of a number of physiographic variables in relationships for predicting the variation of solute loads across a region. This approach has been successfully employed in Australia by Douglas (1973) who found that contrasts in the solute loads from a number of small catchments in Queensland and New South Wales were significantly related to four variables representing a

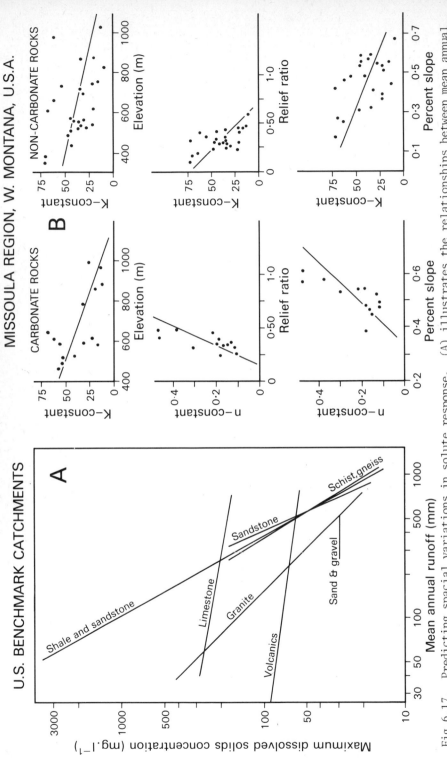

Fig. 6.17. Predicting spacial variations in solute response. (A) illustrates the relationships between mean annual runoff and maximum dissolved solids concentration for catchments developed on specific rock types proposed by Bieseker and Leifeste (1975) from an analysis of data from US bench-mark stations and (B) presents a number of relationships between solute rating curve parameters and catchment characteristics developed by Foggin and Forcier (1977) for streams

162

precipitation seasonality and volume index, mean annual
precipitation, mean annual runoff and the percentage of the
catchment occupied by rain forest. A similar approach was
used by Steele and Jennings (1972) in Texas where significant
relationships were established between a number of solute
parameters and variables indexing catchment area, mean annual
precipitation, average number of thunderstorm days in the year,
mean annual evapotranspiration and annual frequency values of
streamflow at the sampling point. A further refinement was
also introduced into this study by using the residuals from
certain relationships to define regions with their own
individual relationships.

As the scope of a prediction equation is reduced to
encompass a relatively small area, climatic controls will
become progressively less important. Geological contrasts
will often play an important role and Reinson (1976) describes
an analysis of the variation in solute concentration exhibited
by ten tributary basins of the 1950 km^2 Genoa River basin in
southeast Australia which demonstrated a significant
correlation between total dissolved solids concentration and a
geological index (percentage of catchment area occupied by
outcrops of quartz diorite-granodiorite). Work by Foggin and
Forcier (1977) in a 5000 km^2 area around Missoula, Montana,
found that solute levels could be related to both lithological
and topographic characteristics. In this case the dependent
variables were not absolute measures of solute concentration
or load but the constant (K) and the exponent (n) in the simple
solute rating relationship $C = K/Q^n$ for each measuring station.
The individual watersheds were divided into two sample
populations based on basin geology, namely, those developed on
non-carbonate rocks and those developed on carbonate-bearing
strata, and significant relationships were established between
K and n and a number of topographic indices (e.g. Figure
6.17B.). These topographic indices and a geological variable
were also incorporated into regional prediction equations by
using stepwise multiple regression (Table 6.10.). The
physical basis of the relationships obtained were explained in
terms of the K value representing an index of solute
availability and runoff residence time and the n value
reflecting the runoff delivery and storm-period dilution
properties of the basin.

The use of multivariate statistical techniques to develop
predictive equations clearly introduces problems of statistical
rigour, but scope would seem to exist for developing this
approach at both the global and the continental and regional
levels. Nevertheless, significant improvements upon the degree
of explanation achieved by previous studies must probably await
the development of indices of catchment character more closely
related to the physical processes of solute generation. Rock
type, in particular, lacks a sound basis for quantification and
work along the lines of that used to develop indices of
erodibility in sediment yield studies could prove profitable.
In addition, potential exists for use of rapid reconnaissance
surveys to collect single water samples from a number of
locations across an area. These data could be used to 'tune'
prediction techniques which possess rather coarse resolution
or to amplify the more detailed records obtained from a series
of permanent gauging sites (e.g. Walling and Webb, 1978).

Table 6.10.   Regional prediction equations for total solute
              concentrations in streams in the Missoula,
              Montana region.

I   Non-carbonate rocks

    K = 115.8 - 0.096 (elev. x relief ratio) - 32.58 (shape
                                                          ratio)
                                                  $(r^2 = 0.56)$

    n = 9.188 - 0.091 (shape ratio) + 0.181 (relief ratio)
                                                  $(r^2 = 0.19)$

II Carbonate rocks

    K = 1207.7 + 174.4 (% carbonate) - 135.12 (area) +
                                          2.723 (orientation
                                                  $(r^2 = 0.96)$

    n = 3.184 + 0.683 (% slope) - 0.405 (area) + 0.781
                                                  (relief rati
                                                  $(r^2 = 0.44)$

Where: K and n   = values from relationship $C = K/Q^n$

         elev.   = mean elevation above 900 m base level

         relief ratio = local relief/maximum basin length

         shape ratio = basin length/diameter of circle of
                        equal to basin area

         area   = log (area of earth's surface/area of
                        basin)

         orientation = index descriptor of the amount of sola
                        radiation received by a plane
                        representing mean basin aspect and
                        slope

         % carbonate = percentage highly soluble carbonate -
                        bearing parent material

Source: Foggin and Forcier (1977)

164

Modelling of temporal variations in solute yield

The use of a solute concentration or dissolved load rating relationships in conjunction with a continuous streamflow record to estimate the dissolved load of a river also represents one approach to the simulation of temporal variations in solute loads and concentrations. Using this approach, the relatively small number of samples required to define the rating relationship provide the basis for simulating a continuous record. Furthermore, the technique may be used to extend records in time, because the rating relationship may be applied to streamflow records outside the period of sample collection, providing a stationary solute response can be assumed. Steele (1973) presents further details of this modelling strategy in the context of simulating specific conductance records which may in turn be used to predict the total solute concentration and individual ion concentrations. Inherent in this approach, however, are all the limitations of simple concentration/discharge relationships in describing the solute response of a river, some of which have been discussed earlier. Improvements may be introduced by using more complex rating relationships or developing concentration prediction algorithms utilising simple mixing models for individual flow components. For example, Lane (1975) demonstrated that the use of a simple rating relationship with periodic slope and intercept terms improved the degree of explanation of the daily specific conductance records for five streams in the western USA from 54-83% of the variance to 81-87%. O'Connor (1976) provides further discussion of the use of a flow component model to simulate temporal variations in solute response. It must nevertheless be recognised that models involving specific flow components necessitate the separation of the total streamflow hydrograph into these components and that this process may involve arbitrary assumptions.

Where continuous streamflow records are absent, or of short duration, a parametric runoff model such as the Stanford IV watershed model (Crawford and Linsley, 1966) could be used to generate the necessary records, although this step would clearly introduce further errors into the solute load simulation. Hart *et al.* (1964) used a streamflow model of this type as a means of generating data on rates of surface runoff, interflow and groundwater flow from a watershed for use in a predictive relationship involving values of these three runoff components.

The use of streamflow data provides a simple and reasonably effective means of simulating the record of solute concentration and load for most rivers, but its scope is necessarily limited. The runoff processes implicit in the streamflow records cannot provide a complete surrogate for the processes involved in solute generation. Furthermore the approach is essentially empirical and 'black box' and there is little hope of using models of this type to predict changes in solute response associated with a changing hydrometeorological background or catchment condition. Physically-based process models are required for this purpose and, although a general lack of a full understanding of catchment solute processes has hindered their development, some progress towards the goal is evident.

Some attempts to develop models of this latter type have involved the 'pickabacking' of solute simulation subroutines on existing runoff simulation models. Thus Nakamura (1971) describes how specific conductance simulation was incorporate into a Tank type runoff model (Sugawara, 1961). The Wisconsi Hydrologic Transport model developed by Huff (1968) is specifically concerned with modelling contaminant movement through the hydrologic system but possesses many elements of a basic solute model in that atmospheric inputs, vegetal washoff, ion exchange and subsurface movement are involved. In this case the model has been developed from the Stanford I streamflow simulation model (Crawford and Linsley, 1966). The Unified Transport Model (UTM) developed from the Wisconsi model by workers at the Oak Ridge National Laboratory in Tennessee, USA (e.g. Patterson *et al.*, 1975) again does not explicitly set out to model solute yields from watersheds, bu includes sub-models dealing with soil chemical exchange of heavy metals and plant mineral uptake and could form the basi of a process-based solute load model.

Potential for developments in this direction is also evidenced by recent work concerned with the modelling of the quality of irrigation return flows and of nonpoint pollution from agricultural sources. The work of the US Bureau of Reclamation (1977) on the prediction of the mineral quality of irrigation flow may be cited as an example of the former whilst the ACTMO model (Agricultural Chemical Transport Model developed by the US Department of Agriculture (Frere *et al.*, 1975) provides a useful example of the latter.

Modelling solute yields in both time and space

The development of models permitting simulation in both space and time and therefore simulation of detailed solute response at any point over a wide area must constitute a major goal for modelling effort. As in the case of the process-based models discussed above, this goal has not yet been fully achieved, and progress must inevitably await developments in the two constituent branches. However, a number of strategies have been used in this context with some success and two which offer potential for general application may be briefly reviewed.

The first is the approach associated with the nonpoint source mineral water quality model described by Betson and McMaster (1975). In this case the solute response at any point within a river system is characterised by a rating relationship of the form:

$$\text{Concentration} = a \ (\tfrac{Q}{A})^{b}$$

Where: Q = discharge

A = catchment area

a + b = constant and exponent

Use of discharge per unit area instead of absolute discharge allows the equation to be applied throughout the basin without the magnitude of the catchment area and there-fore discharge directly influencing the constant (a).

It has proved possible to develop equations to predict the values of a and b based on catchment characteristics. In their study of the Tennessee Valley, Betson and McMaster (1975) used multiple regression to derive equations for predicting these two coefficients which incorporated measures of forest cover and geology as independent variables viz:

$$a, b, = f \ F, C, S, I, U$$

where F = fraction of watershed that is forested

   C = fraction of watershed developed on carbonate rock

   S = fraction of watershed developed on shale-sandstone rock

   I = fraction of watershed developed on igneous rock

   U = fraction of watershed developed on unconsolidated rock

By using this equation, the rating relationship could be synthesised for any location in the study area from easily measured watershed characteristics. Use of this rating relationship in conjunction with a runoff simulation model would in turn enable the continous record of solute response for that location to be synthesised. Betson and McMaster's study considered the concentration of a number of individual ions as well as those of total dissolved solids, and the efficiency of the model varied according to the solute parameter involved. In the latter case average prediction errors for point values of concentration were $\pm 30\%$.

The approach adopted in this model is essentially empirical and its success in simulating solute loads will rely heavily on both the prediction of the rating relationship and the accuracy of the parallel runoff simulation routine. Furthermore its predictive capacity suffers from all the limitations associated with use of rating relationships for simulation of river loads. Nevertheless, the simplicity of the approach would seem to commend it, and considerable scope for further development could exist if the prediction of the a and b values could be made less empirical.

The second strategy is exemplified by the work of Dixon *et al.* (1970) in developing a hydro-quality simulation model for the Little Bear River in northern Utah. This model permitted simulation of salinity, as represented by specific conductance, throughout the river network. The network was divided into a number of reaches (Figure 6.18A.) and a mass balance model was applied to each of these in order to evaluate the changes in specific conductance in a downstream direction. The inflow to any reach (i) was composed of one or more of six inflow components, namely, outflow from the reach immediately upstream (Q_{i+1}); outflow from river branches tributary to the reach being studied (QBR_i); other natural surface inflow to the reach (QS_i); surface irrigation return flow (QIR_i); groundwater inflow (QGI_i) and municipal and industrial effluent (QEF_i) (Figure 6.18B.).

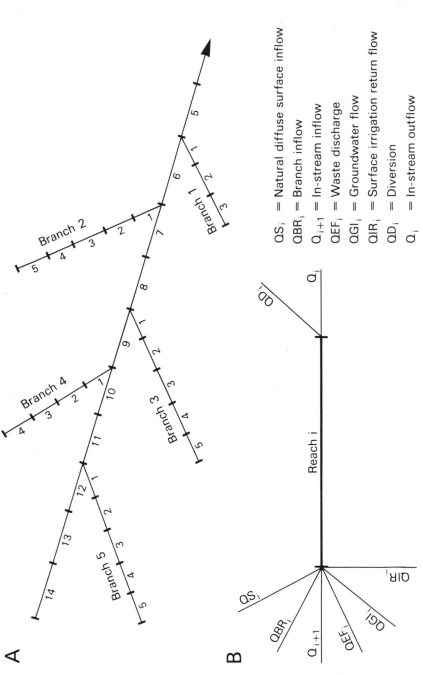

Fig 6.18. Use of a mass balance model to simulate the solute response of a river system. (A) indicates how the network can be divided into a number of reaches and (B) indicates how the inflows and outflows from each of these reaches may be isolated in order to evaluate the mass balance (based on Dixon et al. 1970)

QS_i = Natural diffuse surface inflow
QBR_i = Branch inflow
Q_{i+1} = In-stream inflow
QEF_i = Waste discharge
QGI_i = Groundwater flow
QIR_i = Surface irrigation return flow
QD_i = Diversion
Q_i = In-stream outflow

Each of these inputs was assigned a conductivity (EC) value and the mass balance was applied as:

$$EC_i = (EC_{i+1} \cdot Q_{i+1}) + (ECBR \cdot QBR_i) + (ECS_i \cdot QS_i)$$

$$+ (ECIR_i \cdot QIR_i) + (ECGI_i \cdot QGI_i)$$

$$+ ECEF_i \cdot QEF_i) \Big/ Q_i + QD_i$$

where Q_i = Total in-stream outflow from study reach

QD_i = Diversions from study reach

EC_i = Specific conductance of study reach

A hydrologic submodel was used to generate the necessary water inflows on a monthly basis and the conductivity values associated with the individual inflow components were estimated using a number of different procedures. EC_{i+1} was the outflow conductivity of the upstream reach, $ECBR_i$ and ECS_i were estimated from generalised rating relationships based on available data for the study area, $ECIR_i$ was based on an arbitrary multiplier applied to the conductivity of irrigation supply water, ECG_i was estimated from the conductivity of local groundwater samples and $ECEF_i$ was estimated according to the nature of the effluent discharging activity.

Again, therefore, this model relies heavily upon empirical data, but the mass balance approach could be used as basis for developing a more sophisticated model of the solute response of a river network if more process-based procedures could be substituted to predict the conductivities associated with individual inflow components, and if the time base of the simulation could be reduced to afford increased temporal resolution.

Overall, the modelling of river solute loads on both a spatial and temporal basis is a field that requires further development of process-based procedures for predicting solute response. Empirical procedures have provided the basis for worthwhile initial developments but future progress must involve simulation of such mechanisms as the release of solutes by weathering, plant - solute interactions, soil water residence times and in-transit sediment-solute interaction. Furthermore, these mechanisms must be analysed in the context of recent knowledge of runoff forming processes and the various implications of the partial area and variable source-area concepts of runoff generation (e.g. Kirkby, 1978). It may be insuffiecient to append solute-orientated subroutines onto runoff models which take little account of these concepts.

REFERENCES

Alekin, O.A., and Braznikhova, L.V., 1960, A contribution on runoff of dissolved substances on the world's continental surface, *Gidrochim. Mat.*, 32, 12-34.

American Public Health Association, 1965, *Standard methods for the examination of water and waste water*, (A.P.H.A., New York)

Anderson, P.W., 1963, Variations in the chemical character of the Susquehanna River at Harrisburg, Pennsylvania, *US Geological Survey Water Supply Paper* 1779 B.

Archer, R.J., La Sala, A.M., and Kammerer, J.C., 1968, Chemical quality of streams in the Erie-Niagara Basin, New York, *New York State Water Resources Commission Basin Planning Report* ENB-4.

Arnett, R.R., 1978, Regional disparities in the denudation rate of organic sediments, *Zeitschrift fur Geomorphologie*, Supplementband 29, 169-179.

Bestow, T.T., 1977, The movement and changes in concentration of contaminants below a sanitary landfill, Perth, Western Australia, in *Effects of urbanization on the hydrological regime and on water quality*, IAHS Publication No. 123, 370-379.

Betson, R.P., 1964, What is watershed runoff? *Journal of Geophysical Research*, 69, 1541-51.

Betson, R.P., and McMaster, W.M., 1975, Nonpoint source mineral water quality model, *Journal, Water Pollution Control Federation*, 47, 2461-2473.

Biesecker, J.E., and Leifeste, D.K., 1975, Water quality of hydrologic bench marks - an indicator of water quality in the natural environment, *US Geological Survey Circular* 460E.

Blanc, P., and Conrad, G., 1968, Evolution geochimique des eaux de l'Oued Saoura (Sahara nord-occidental), *Rev. Geogr. Phys. Geol. Dyn.*, 10, 415-428.

Brinson, M.M., 1976, Organic matter losses from four watersheds in the humid tropics. *Limnology and Oceanography* 21, 572-582.

Brown, E., Skougstad, M.W., and Fishman, M.J., 1970, Methods for collection and analysis of water samples for dissolved minerals and gases. *Techniques of Water-Resources Investigations of the United States Geological Survey*, Book 5, Chapter A1.

Claridge, G.G.C., 1970, Studies in element balance in a small catchment at Taita, New Zealand, *IAHS Publication*, No. 96, 523-40.

Collier, C.R. *et al.*, 1970, Influences of strip mining on the hydrological environment of parts of Beaver Creek Basin, Kentucky, 1955-1966, *US Geological Survey Professional Paper* 427C.

Corbel, J., 1964, l'erosion terrestre, etude quantitative, *Annales de Geographie*, 73, 385-412.

Crawford, N.H., and Linsley, R.K., 1966, *The Stanford Watershed Model Mk. IV.*,, *Tech. Rept.*, 39, Dept. Civil Engineering, Stanford University.

Dixon, M.P., Hendricks, D.W., Huber, A.L., and Bagley, J.M., 1970, *Developing a hydro-quality simulation model*, *Tech. Rept.* PRWG 67-1, Utah State University.

Dobbs, R.A., Wise, R.H., and Dean, R.B., 1972, The use of ultraviolet absorbance for monitoring the total organic carbon content of water and wastewater, *Water Research*, 6, 1173-1180.

Douglas, I., 1972, The geographical interpretation of river water quality data, *Progress in Geography*, 4, 1-81.

Douglas, I., 1973, Rates of denudation in selected small catchments in Eastern Australia, *University of Hull Occasional Papers in Geography*, 21.

Durum W.H., 1953, Relationship of the mineral constituents in solution to stream flow, Saline River near Russell, Kansas, *Transactions, American Geophysical Union*, 34, 435-442.

Durum, W.H., Heidel, S.G., and Tison, L.J., 1960, World-wide runoff of dissolved solids, *IAHS Publication* No. 51, 618-28.

Ellis, J.B., 1975, Urban stormwater pollution. *Middlesex Polytechnic Research Report*.

Environment Canada, 1976, *Lake Erie in the early seventies*, Special issue of *Journal, Fisheries Research Board of Canada*, 33, No. 3.

Finlayson, B., 1977, Runoff contributing areas and erosion, *University of Oxford, School of Geography, Research Papers* No. 18.

Foggin, G.T., and Forcier, L.K., 1977, Using topographic characteristics to predict total solute concentrations in streams draining small forested watersheds in Western Montana, *University of Montana Joint Water Resources Research Center Rept.*, No. 89.

Frere, M.H., Onstad, C.A., and Holtan, H.N., 1975, ACTMO, an agricultural chemical transport model, *US Department of Agriculture Rept.* ARS-H-3.

171

Froehlich, W., 1975, Dynamika transportu fluwialnego Kamienicy Nawojowskiej. *Polska Academia Nauk Instyt Geografii i Przestrzennego Zagospodarowania Prace Geograficzne* Nr. 114.

Gibbs, R., 1970, Mechanisms controlling world water chemistr *Science*, 170, 1088-1090.

Glover, B.J., and Johnson, P., 1974, Variations in the natur chemical concentration of river water during flood flows, and the lag effect, *Journal of Hydrology*, 22, 303-16.

Gorham, E., 1961, Factors influencing supply of major ions t inland waters with special reference to the atmosphe *Geological Society of America Bulletin*, 72, 795-840.

Green, D.B., Logan, T.J., and Smeck, N.E., 1978, Phosphate adsorption-desorption characteristics of suspended sediments in the Maumee River basin of Ohio. *Journc of Environmental Quality*, 7, 208-12.

Gunnerson, C.G., 1967, Streamflow and quality in the Columbi River basin, *Proceedings, American Society of Civil Engineers, Journal of the Sanitary Engineering Divis* 93, 5626-5636.

Hall, F.R., 1970, Dissolved solids-discharge relationships. 1. Mixing models, *Water Resources Research*, 6, 845-5

Hall, F.R., 1971, Dissolved solids-discharge relationships. 2. Applications to field data, *Water Resources Resec* 7, 591-601.

Harned, H.S., and Owen, B.B., 1958, *The physical chemistry c electrolytic solutions*, (Reinhold, New York).

Hart, F.C., King, P.H., and Tchobanoglous, G., 1964, Discuss of 'Predictive techniques for water quality inorgani by J.R Ledbetter and E.F. Gloyna, *Proceedings, American Society of Civil Engineers, Journal of the Sanitary Engineering Division*, 90, 63-64.

Hem, J.D., 1970, Study and interpretation of the chemical characteristics of natural water, *US Geological Suru Water Supply Paper* 1473.

Hendrickson, G.E., and Krieger, R.A., 1964, Geochemistry of natural waters of the Blue Grass Region, Kentucky, *l Geological Survey Water Supply Paper* 1700.

Hewlett, J.D., and Hibbert, A.R., 1967, Factors affecting tl response of small watersheds to precipitation in hum areas, in *International Symposium on Forest Hydroloc* (Pergamon, Oxford) 275-90.

Hilder, D.W., and Wilson, A.L., 1972, Statistical tests required in checking the homogeneity of impurities in cross-sections of rivers, *Water Research Association Technical Memorandum*, T.M. 72.

Holeman, J.N., 1968, The sediment yield of major rivers of the world, *Water Resources Research*, 4, 737-47.

Hotes, F.L., and Pearson, E.A., 1977, Effects of irrigation on water quality, in *Arid land irrigation in developing countries: Environmental problems and effects*, ed Worthington, E.B., (Pergamon, Oxford) 127-58.

Huff, D.D., 1968, *Simulation of the hydrologic transport of radioactive aerosols*. Ph.D. Dissertation, Committee on Hydrology, Stanford University.

Hutchinson, G.E., 1937, Limnological studies in Indian Tibet, *Internat. Rev. der gesamten Hydrobiol. und Hydrog.*, 35, 134-75.

Janda, R.J., 1971, An evaluation of procedures used in computing chemical denudation rates, *Geological Society of America Bulletin*, 82, 67-80.

Jaworska, M., 1968, Erozja chemiczna i dendacja zlewni rzek Wieprza i Pilicy, *Prace Panstwowego Instytutu Hydrologiczno-Meteorologicznego*, 95, 29-47.

Jaworski, N.A., and Hetling, L.J., 1970, Relative contributions of nutrients to the Potomac River basin from various sources, in *Proceedings, 1970 Cornell Agricultural Waste Management Conference*, 134-46.

Johnson, A.H., Bouldin, D.R., Goyette, E.A., and Hedges, A.M., 1976, Phosphorus loss by stream transport from a rural watershed: Quantities, processes and sources, *Journal of Environmental Quality*, 5, 148-157.

Johnson, N.M., Likens, G.E., Bormann, F.H., Fisher, D.W., and Pierce, R.S., 1969, A working model for the variation in stream water chemistry at the Hubbard Brook Experimental Forest, New Hampshire, *Water Resources Research*, 5, 1353-63.

Judson, S., and Ritter, D.F., 1964, Rates of regional denudation in the United States, *Journal of Geophysical Research*, 69, 3395-401.

Kirkby, M.J.(ed), 1978, *Hillslope hydrology*, (Wiley Interscience, Chichester).

Knedlhans, S., 1971, Mechanical sampler for determining the water quality of ephemeral streams, *Water Resources Research*, 7, 728-30.

Kunkle, S.H., 1972, Effects of road salt on a Vermont stream, *Journal, American Waterworks Association*, 64, 290-4.

Lane, W.L., 1975, Extraction of information on inorganic water quality, *Hydrology Papers, Colorado State University* No. 73.

Langbein, W.B., and Dawdy, D.R., 1964, Occurrence of dissolved solids in surface waters in the United States, *US Geological Survey Professional Paper* 501D, D115-D117.

Langbein, W.B., and Schumm, S.A., 1958, Yield of sediment in relation to mean annual precipitation, *Transactions, American Geophysical Union*, 39, 1076-84.

Ledbetter, J.O., and Gloyna, E.F., 1964, Predictive technique for water quality inorganics, *Proceedings, American Society of Civil Engineers, Journal of the Sanitary Engineering Division*, 90, 127-151.

Likens, G.E., Bormann, F.H., Johnson, N.M., and Pierce, R.S. 1967, The calcium, magnesium, potassium and sodium budgets for a small forested ecosystem, *Ecology*, 48, 772-85.

Livingstone, D.A., 1963a, Chemical composition of rivers and lakes. Data of geochemistry, *US Geological Survey Professional Paper* 440G.

Livingstone, D.A., 1963b, The sodium cycle and the age of the ocean, *Geochim. Cosmochim. Acta*, 27, 1055-69.

Loughran, R.J., and Malone, K.J., 1976, Variations in some stream solutes in a Hunter Valley catchment, *Research Papers in Geography, University of Newcastle, N.S.W.*

Maciolek, J.A., 1962, Limnological organic analysis by quantitative wet oxidation, *US Fish and Wildlife Service Rept.* 60.

Malcolm, R.L., and Durum, W.H., 1976, Organic carbon and nitrogen concentrations and annual organic carbon load of six selected rivers of the United States, *US Geological Survey Water Supply Paper* 1817-F.

Mattson, J.S., Smith, C.A., Jones, T.T., Gerchakov, S.M., and Epstein, B.D., 1974, Continuous monitoring of dissolved organic matter by ultra-violet visible photometry, *Limnology and Oceanography*, 19, 530-35.

Matveyev, A.A., and Bashmakova, O.I., 1967, Chemical composition of atmospheric precipitation in some regions of the USSR, *Soviet Hydrology*, 480-91.

McElroy, A.D., Chiu, S.Y., Nebgen, J.W., Aleti, A., and Bennett, F.W., 1976, Loading functions for assessment of water pollution from nonpoint sources, *US Environmental Protection Agency Environmental Protection Technology Series Report* EPA-600/2-76-151.

Meade, R.H., 1969, Errors in using modern stream-load data to estimate natural rates of denudation, *Geological Society of America Bulletin*, 80, 1265-74.

Menzel, D.W., and Vaccaro, R.F., 1964, Measurement of dissolved organic and particulate carbon in sea water, *Limnology and Oceanography*, 9, 138-42.

Meybeck, M., 1976, Total mineral dissolved transport by world major rivers, *Hydrological Sciences Bulletin*, 21, 265-84.

Miller, C.R., 1951, Analysis of flow duration sediment rating curve method of computing sediment yield, *US Bureau of Reclamation Rept.*

Miller, J.P., 1961, Solutes in small streams draining single rock types, Sangre de Cristo Range, New Mexico, *US Geological Survey Water Supply Paper* 1535-F.

Nakamura, R., 1971, Runoff analysis by electrical conductance of water, *Journal of Hydrology*, 14, 197-212.

Nelson, D.J., 1970, Measurement and sampling of outputs from watersheds, in *Analysis of temperate forest ecosystems*, ed. Reichle, D.E., 242-285.

O'Connor, D.J., 1976, The concentration of dissolved solids and river flow, *Water Resources Research*, 12, 279-94.

Omernik, J.M., 1976, The influence of land use on stream nutrient levels, *US Environmental Protection Agency, Ecological Research Series Report* EPA-600/3-76-014.

Patterson, M.R., Munro, J.K., Fields, D.E., Ellison, R.D., Brooks, A.A., and Huff, D.D., 1974, A users manual for the FORTRAN IV version of the Wisconsin Hydrologic Transport Model, *Oak Ridge National Laboratory Report* ORNL/NSF/EATC-7.

Pavelko, I.M., and Tarasov, M.N., 1967, Hydrochemical maps of the rivers of Kazakhstan and their use for the rapid forecasting of the mineralization and ion composition of waters of prospective reservoirs, *Soviet Hydrology*, 495-507.

Peck, A.J., 1976, Interaction between vegetation and stream water quality in Australia. *Proceedings of the Fifth Workshop of the United States/Australia Rangelands Panel*, 149-55.

Pierce, R.S., Hornbeck, J.W., Likens, G.E., and Bormann, F.I 1970, Effects of vegetation elimination on stream water quantity and quality, in *Results of research* representative and experimental basins. *Proceeding* of the Wellington Symposium IAHS Publication, 96, 311-28.

Pionke, H.B., Nicks, A.D., and Schoof, R.R., 1972, Estimatii salinity of streams in Southwestern United States, *Water Resources Research*, 8, 1597-1604.

Rainwater, F.H., 1962, Stream composition of the contermino United States, *US Geological Survey Hyd. Inv. Atlas* HA 61.

Rainwater, F.H., and Thatcher, L.L., 1960, Methods for collection and analysis of water samples, *US Geological Survey Water Supply Paper* 1454.

Reinson, G.E., 1976, Hydrogeochemistry of the Genoa River basin, New South Wales - Victoria. *Australian Journ* of Marine and Freshwater Research, 27, 165-186.

Robbins, J.W.D., Howells, D.H., and Kriz, G.J., 1972, Strear pollution from animal production units, *Journal, Wa* Pollution Control Federation, 44, 1536-44.

Skakalskiy, B.G., 1966, Basic geographical and hydrochemica: characteristics of the local runoff of natural zone: in the European territory of the USSR, *Transactions* State Hydrological Institute (Trudy GGI) 137, 125-8(

Smart, P.L., Finlayson, B.L., Rylands, W.D., and Ball, C.M. 1976, The relation of fluorescence to dissolved organic carbon in surface water, *Water Research*, 10 805-811.

Steele, T.D., 1969, Digital-computer applications in chemicz -quality studies of surface water in a small watersh *IAHS Publication*, 80, 203-14.

Steele, T.D., 1973, Simulation of major inorganic chemical concentrations and loads in streamflow, *US Geologice* Survey Computer Contribution.

Steele, T.D., and Gilroy, E.J., 1971, Statistical technique: for assessing longterm changes in streamflow salini* *Transactions, American Geophysical Union*, 52, 846.

Steele, T.D., and Jennings, M.E., 1972, Regional analysis o: streamflow chemical quality in Texas, *Water Resource* Research, 8, 460-77.

Strakhov, N.M., 1967, *Principles of Lithogenesis* vol. 1, (Oliver and Boyd, Edinburgh).

Sugawara, M., 1961, On the analysis of runoff structure about several Japanese Rivers, *Japanese Journal of Geophysics*, 2, 1-77.

Toler, L.G., 1965, Relation between chemical quality and water discharge in Spring Creek, Southwestern Georgia, *US Geological Survey Professional Paper* 525C, 209-13.

US Bureau of Reclamation, 1977, Prediction of mineral quality of irrigation return flow, *US Environmental Protection Agency Report* EPA/600/2-77/179.

Voronkov, P.P., 1963, Hydrochemical bases of the separation of local runoff and a method of separating its discharge hydrograph, *Meteorologiya i Gidrologiya*, 8, 21-8.

Walling, D.E., 1974, Suspended sediment and solute yields from a small catchment prior to urbanization. in *Fluvial processes in instrumented watersheds*, ed. Gregory, K.J., and Walling, D.E., *Institute of British Geographers Special Publication* No. 6, 169-192.

Walling, D.E., 1975, Solute variations in small catchment streams: some comments. *Transactions, Institute of British Geographers*, 64, 141-7.

Walling, D.E., 1978, Reliability considerations in the evaluation and analysis of river loads, *Zeitschrift fur Geomorphologie*, Supplement Band 29, 29-42.

Walling, D.E., and Webb, B.W., 1975, Spatial variation of river water quality: a survey of the River Exe, *Transactions, Institute of British Geographers*, 65, 155-69.

Walling, D.E., and Webb, B.W., 1978, Mapping solute loadings in an area of Devon, England, *Earth Surface Processes* 3, 85-99.

Wartiovaara, J., 1975, Jokien ainevirtaamista Suomen rannikolla. *Vesientutkimuslaitoken Julkaisuja*, 13.

Wilcox, L.V., 1962, Salinity caused by irrigation, *Journal, American Waterworks Association*, 54, 217-22.

Zverev, V.P., 1971, Hydrochemical balance of the USSR territory, *Dokl. Akad. Nauk. SSSR*, 198, 161-3.

7.

Measurement of reservoir sedimentation

D. L. Rausch and H. G. Heinemann

Introduction

A reservoir sedimentation survey is an excellent method to determine the sediment yield from a watershed. This method has distinct advantages over other approaches. The measurements obtained are generally much more accurate than those obtained using sediment yield prediction procedures. In contrast with the streamflow measuring and suspended sediment sampling method, the sediment yield for past years can be determined with one survey--without waiting years for collection of streamflow data. The sedimentation survey can be made at a convenient time, whereas streamflow stations generally must be attended by stream gaugers periodically and whenever flow events occur, day or night. This method is usually cheaper than others, and once the ranges and the benchmarks are established, a reservoir can be resurveyed in only a few days.

There is a limitation to the type of survey where thickness of sediment deposition is determined with a spud or piston sampler. For reservoirs built on sandy sites and having sandy inflows, distinguishing between deposits and the original bottom in the upstream reservoir area is often difficult, if not impossible. Consequently, the original reservoir capacity cannot be computed accurately. Other data needed to determine sediment yield from reservoir surveys include the history of the reservoir, when the reservoir started to fill, and whether any sediment has been removed. Also, the sediment discharged through the spillways must be estimated. This can be done with trap efficiency values determined from Brune's (1953) or Heinemann's (1981) procedure.

This chapter describes an efficient method for making a good reservoir sedimentation survey and the procedure for determining the sediment yield from a watershed. The suggested method is based on considerable experience in making sedimentation surveys of small reservoirs up to 600 m wide and 10 m deep. Equipment and manpower needs, safety precautions, computational methods, suggested analyses of sediment samples,

Agricultural Engineer, Watershed Research Unit, ARS, USDA, Columbia, Missouri, and Hydraulic Engineer (deceased), Hydrology lab., SEA-AR, USDA, Beltsville, Maryland.

and other components to complete a sedimentation survey are described.

Another reference on this subject is the American Society of Civil Engineers Sedimentation Manual No. 54 (Vanoni, 1977) which covers sediment deposit measurement techniques and equipment used for both large and small reservoirs. Information for surveying small agricultural reservoirs is contained in articles by Rausch and Heinemann (1968); Heinemann and Rausch (1971); UDSA, SCS's National Engineering Handbook on Sedimentation (1973); and Rausch and Heinemann (1976).

Publications dealing primarily with the gamma probe for determining the sediment volume-weight include those by Timblin and Florey (1957), Heinemann (1962), McHenry (1962), McHenry and Dendy (1964), and by McHenry *et al.* (1971). One publication, Heinemann and Dvorak (1965), deals primarily with the volume computations.

General approach

Each survey should obtain enough data to accurately determine sedimentation rates with the most efficient surveying method or methods. In addition, the field survey should be documented with benchmarks to expedite future surveys.

Fieldwork

The fieldwork required depends upon the choice of surveying methods. It may include aerial and topographic mapping, locating ranges, cross sectioning with a level (above water) and by sounding (below water), and in situ measurement and sampling of sediment deposits. It is highly desirable that the original survey be made soon after construction is completed and a topographic map constructed. If a topographic map is not available, an accurate map must be constructed, because any error in the calculation of the original reservoir capacity will cause an equal error in the computed sediment volume.

Before starting the survey work, the topography in the surrounding area should be studied, including that directly downstream from the dam. The slopes leading into the reservoir should be examined to determine if they are continuous or if bench terraces or other irregularities exist. Such topographic features require adequate delineation during the fieldwork. Also, the method for computing the reservoir volume must be chosen so that compatible field data can be collected.

Surveying and computational methods

Three basic surveying methods, outlined by Eakin and Brown (1939), are (1) contour, (2) cross-sectional area or range, and (3) a combination of these two. In most surveys, a combination of the contour and cross-sectional area methods is used.

The contour survey method provides adequate information for the following methods of computing reservoir capacities as outlined by Heinemann and Dvorak (1965):

1. Stage-area curve.

2. Modified prismoidal formula.

3. Simpson's rule using contour areas.

4. Average-contour area formula.

The cross-sectional area survey or range method provides adequate data for the following methods of calculating reservoir capacities:

1. Eakins range end formula.

2. Cross-sectional area versus distance from dam curve.

3. Simpson's rule using cross-sectional areas.

4. Average-end area formula.

Care must be exercised when using an average-end area formula. A sizeable volumetric error usually occurs when there is a considerable size difference between the two areas being averaged. The error, however, is minimal if the areas are similar in size, have a similar width or depth dimension, or the reservoir segment is bowl-shaped.

The combination survey method is used advantageously on many reservoirs. The main body of the reservoir can usually be surveyed best by the cross-sectional or range method, because the topography is more uniform and its shape changes less between surveys than in deltaic areas. In contrast, the topography in exposed deltaic areas may not be uniform, or it may change considerably from survey to survey and it can best be surveyed by the contour method. This combination method is flexible and easily adapted to most reservoirs.

Accuracy

The accuracy needed in surveying sediment volume and density depends upon the intended use of the finding and the accuracy of the associated data. Changes in sediment volume between surveys are affected by sediment yield, average trap efficiency (TE) of the reservoir, removal of sediment, and/or change in the water level. Sediment yield and TE are not constants. They are affected by meteorological events, land use, and other conservation practices in the watershed. The sediment deposition rate cannot be accurately computed unless all the above data are available or can be closely estimated. Thus, if these data cannot be determined quite accurately, the survey itself may not need to be very accurate. However, if all sediment outflow has been measured since dam construction or a previous survey, then an accurate survey will also give an accurate measure of TE and sediment yield for the period.

Horizonal control of ±0.5 m on any area exceeding 100 m square will give an areal accuracy better than ±1%. Ground and sediment surface elevations are usually measured to within ±3 cm and benchmarks are surveyed to within ±0.3 cm.

181

For a reservoir that averages 2 m of water depth and 1 m of sediment depth, this gives an accuracy of $\pm$1.5% for the water volume and $\pm$3% for the sediment volume. These accuracies are not hard to obtain, and as areas and depths increase, the accuracies can be improved.

Manpower and communications requirements

    Reservoir surveys can usually be divided into two parts: (1) land survey, and (2) water survey and sediment sampling. Each can be performed efficiently by a 2- or 3-man crew, either simultaneously or by two crews or sequentially by the same crew.
    Small, hand-held, two-way radios speed the performance of crews, especially a new crew on a larger reservoir. They are also helpful on rough terrain where the instrument man can see the level rod but not the rodman.

Crew safety

    Reservoir surveys present diverse hazards to each participant. Working in the boats can be hazardous. Walking surfaces on the boats should be non-skid, and a good life jacket should be worn at all times. Reflection off the water in addition to the direct sunlight makes the effect of the sun more intense and measures should be taken to protect personnel from sunburn, heat exhaustion, and stroke. In general, use common sense and caution in surveys on and about the water.
    Poisonous snakes and insects are sometimes more numerous in or near bodies of water, and crew members should be aware of the hazards and take appropriate precautions. Snake bite kits should be available when working in areas too remote for medical help. Personnel allergic to insect stings should carry their appropriate medicine.
    If a gamma density probe is used on the survey, personnel should be trained in its proper use. They should be required to wear film badges that record accumulated radiation received or dosimeters that given an immediate reading of accumulated radiation, or both.

Surveying equipment

Boats

    Two 4.2 m (or longer) aluminium boats with motors are used as a means of transportation and as a work platform when fastened together by two 5 cm x 25 cm x 4 m planks, as shown in Figures 7.1. and 7.2. Appropriate rafts might be substituted for the boats.

Cable and reel

    To make sure that all measurements along a range across the reservoir are on line, a cable is stretched along the range. The cable is a 3 mm, 7- by 19-strand, galvanized, preformed aircraft cable or equivalent. The cable is stored

Fig 7.1. and 7.2.   Side and front views of two boats
fastened together by planks and equipped for reservoir
sedimentation surveys.

and tightened by the reel shown in Figure 7.3.   This reel
will hold 600 m of cable.

Cable distance measurements

    The distance across the reservoir is measured by using
a commercially available cable-measuring meter with smooth,
hardened-steel wheels that ride the cable.   Similar devices
may be constructed of various homemade designs.   For small
reservoirs a simple measuring tape or chain can be used, but
a cable is much more convenient to handle.

Water depth

    Water depth can be measured manually with a long,
neutrally buoyant pole that is marked for length and which
has a 12 cm disc on the bottom, or with a simple 2 kg
sounding bell that is 12 cm in diameter and is attached to a

Fig 7.3. Reel holding cable which is stretched across reservoir to guide boat.

calibrated rope. These simple devices and techniques are recommended for small reservoirs (ponds). For larger reservoirs, however, use of a sonic depth recorder may be desirable. Depth recorders determine the depth of water by measuring the elapsed time between sending a high-energy acoustic signal downward and receiving the reflected signal off the bottom. Recorder adjustments permit 'zeroing' the chart at the water surface and varying the depth scale to compensate for differences in the speed of sound in the water.

Typical limits for the depth recorder are from 0.5 to 54 m. Depths as shallow as 0.5 m can be measured if the outboard transducer, or 'fish', is mounted 15 cm below the water surface instead of the recommended 60 cm (Figure 7.4.) The water depth versus time is recorded on a chart. Distance across the reservoir can also be recorded on the same chart if a switch is installed on the cable meter (Figure 7.5.) and wired parallel to the 'fix' switch on the recorder. The switch closes momentarily for every 'set' distance of boat travel and causes a vertical line to be marked on the chart (Figure 7.6.). This system has performed satisfactorily at boat speed up to 0.6 m.

Survey procedure

Base line

To provide accurate horizontal control for the survey, a base line is established along on side and approximately parallel to the reservoir (Figure 7.7.). Three or four permanent markers along this line enhance survey accuracy. Sometimes the base line has to be located on top of the dam. Steel pipes or rods driven to below the frost line and flush with the ground, or concrete monuments, serve as good permanent markers. They should be reflected by distance and direction to nearby permanent objects.

Fig 7.4.   Mounting of sonic depth recorder transducer.

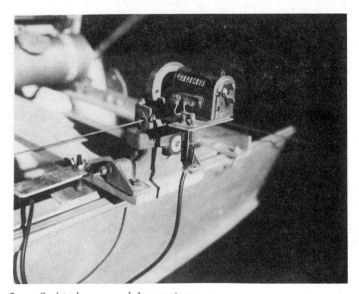

Fig 7.5.   Switch on cable meter.

Benchmarks

At least two permanent elevation benchmarks should be
established during the survey.   This may include a designated
point on a concrete spillway or similar object.   It is
sometimes convenient to have other benchmarks around the lake
facilitate and check the level work.

Contour surveying

The contour surveying of a reservoir requires numerous elevation determinations of the reservoir topography and accurate location of these points so these can be plotted. The number of soundings or elevations needed varies with the irregularity of the topography. Various methods and equipme can be used to loate the elevation points such as transit, plane table and alidade, or range number and distance. A number of patterns can be used to systematically cover the area with elevations, such as a grid system or a certain distance between points along ranges. When an adequate number of specific point elevations are obtained and plotted contour lines can be drawn to reflect the reservoir topograp

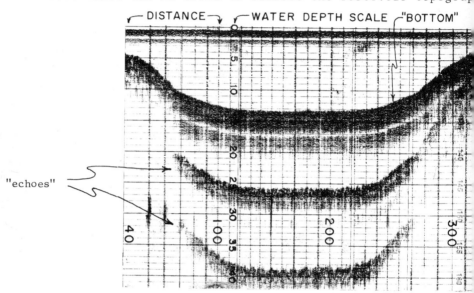

Fig 7.6. Depth recorder chart showing distance and water depth.

Experience has shown that a good reservoir contour map can be drawn from a field determination of the upper spillwa contour, a selected lower contour such as the water surface elevation, and cross-sectional data from a number of properl located ranges. Information on the upper (emergency) spillw contour and a lower contour provide excellent areal control. The best elevation for the lower contour is about one-third of the total reservoir depth below the upper spillway. Both contours provide much information on the topography between ranges, and data from these two contours provide strategic points on a stage area curve. The plane table and alidade may be used to map these selected contours and range ends and to accurately define the volume of exposed deltaic deposits. Maintaining vertical control with an engineer's level while locating the points with the alidade will speed the plane table work and enhance its accuacy. If the reservoir has been surveyed previously and the shoreline has not changed, only the deltas need to be resurveyed by the plane table method. Experience has shown that a 0.5 m contour interval is usually adequate for small reservoirs. However, additional contour lines may be needed in the areas to accurately define this sediment deposit.

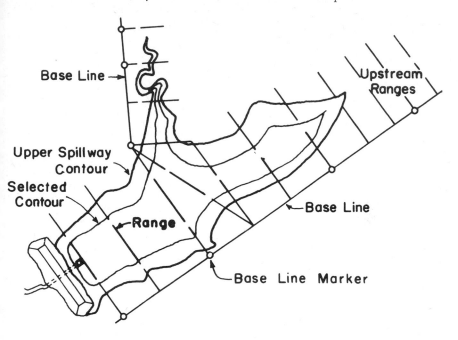

Fig 7.7.   Suggested range layout for a typical reservoir.

Range layout

     Cross sections or ranges should be placed generally perpendicular to the valley, and, if possible, perpendicular to the base line and parallel to each other because this speeds the survey and computations. Although such placement is not always possible, the location of ranges must always be selected carefully so that the data obtained along them accurately describe the topography. A suggested range layout is shown in Figure 7.7. Ranges on a small pond may be spaced as closely as 15 m. As the reservoir size increases, the interval between ranges may also increase without losing accuracy. For a reservoir with a 60 ha surface area, the spacing may be up to 150 m if the topography is uniform. The reservoir segment between any two ranges, however, should not be a disportionate part (more than 25%) of the reservoir capacity. Ranges must be locted so that they will yield maximum data for drawing contours. It is better to obtain more than adequate, rather than insufficient, data. Ranges should be extended if necessary, so that maximum elevations are at least 1 m above the highest spillway. If the exact topography of the upstream face of the dam is not known, at least one range should be located perpendicular to the axis of the dam to establish the presence or absence of a berm

187

and also to locate the toe of the fill. If only a minimum
number of ranges is used, one range must be located across
the reservoir along the upstream toe of the dam.
   When the range layout has been decided, the ends of the
ranges are marked either temporarily or permanently and
located by survey. Steel range markers driven flush with th
ground can easily be located with a metal detector during
succeeding surveys

Range surveying

   Range cross sections are surveyed in the conventional
manner, beginning at the base line or convenient range end
and proceeding towards the water. If the base line is not
above the upper spillway contour, the range line is extendec
in the opposite direction and cross sectioned until it is 1
above the upper contour. When the range has been surveyed
down to the water, a stake is set at the water line and marl
with the distance from the base line or range end. This
distance is used to set the cable meter when the water porti
of the range is surveyed from the boat. A similar stake is
set on the opposite shore and marked with the distance basec
on the cable meter reading. The rest of the range is survey
to 1 m above the upper contour elevation or to range end.

Water depth

   Water depths can be measured manually with a sounding b
or pole, or automatically with a depth recorder. To guide 1
boat from which measurements are made, the cable is threadec
through the cable meter and stretched across the reservoir.
The boat containing the cable meter and depth recorder rema
near the cable reel while the other boat pulls the cable ac
the reservoir and anchors it on the other shore.
   Be cautious when tightening the cable. People should
stand clear in case the cable breaks or the end anchors pul
loose. Also, when loosening the cable, the operator should
maintain firm control of the reel crank to prevent it from
spinning free and possibly causing injury. Others on the la
should be warned that the cable is stretched across the
reservoir so they will not strike it with their boats and mc
   When the cable is tight and the cable meter is set with
the distance from the range end, the boat motor and depth
recorder are started and the water depths across the reserve
are recorded. The distance lines are marked on the chart a
it is being produced, as shown in Figure 7.6. The areas of
each range too shallow for the depth recorder are sounded
manually and recorded.
   The notes should include range station or distance from
range initial point, water depth, and depth to original bot
(initial survey only). Water level is recorded two or thre
times daily during the survey and referenced to the bench m
to provide good vertical control for range cross sectioning
Water depths can then be converted to elevations and plotte

Sounding reservoir original bottom

   If the original bottom has not yet been surveyed previo
one of the following tools can be used to determine the
original depth: a spud (Figure 7.8.), a smooth sounding pol

Fig 7.8. Sectional spud used to penetrate sediment and
determine original bottom elevations, shown with meter stick
and adapter for extension pipe.

a piston sampler, or any combination of these. Because the
original soil surface is usually more cohesive than the
sediment, it will adhere better to the triangular grooves in
the spud. The spud is vertically into the sediment deposits
and retrieved from deep water by the attached rope or from
shallow water by an extension pipe.

When the sounding pole and other probes and samplers reach
the original soil, penetration resistance usually increases
suddenly, thus indicating the location of the original bottom.
The sediment sampler may be used to obtain a sample of the
original bottom, and the interface between sediment and
original soil can be determined based on composition,
structure, degree of aggregation, colour, and coarseness of
material and accumulation of leaves, twigs and other organic
matter. The easiest method is to use the sounding pole and
periodically check the results with the spud or sediment
sampler. By having distances marked on the sounding pole and
other devices, one can easily convert a given depth below the
water surface to an elevation from the elevation of the water
surface. The original bottom shoul be probed about as often
as the sediment surface is sounded, or, if, a depth recorder
is used, about every 3 to 7.5 m, depending on range length
and smoothness of the bottom.

It is sometimes quite difficult and strenuous to determine
the original bottom of a reservoir. Therefore it is wise to
survey reservoirs that may be used in sedimentation studies
right after construction. Furthermore, ranges can be located
and surveyed to provide maximum topographic data if the survey
is made before the reservoir is filled. In addition, breaks
in slopes, borrow areas, and other irregularities can be
surveyed more easily and accurately.

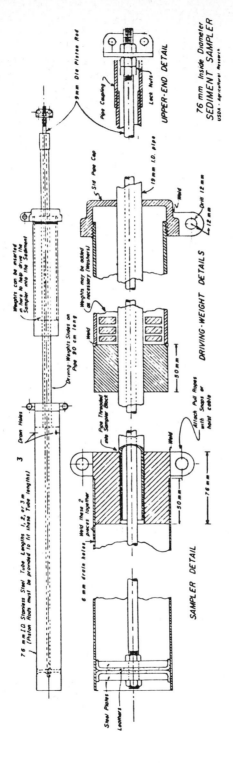

Fig 7.9. Drawing of ARS volumetric sediment sampler which is made out of standard plumbing pipe materials (Inside diameter is 7.6 cm).

SEDIMENT MEASURING EQUIPMENT

Sediment sampler

A piston-type sampler is used to take undisturbed volumetric samples of the deposited sediment. Figure 7.9. is a drawing of the Agricultural Research Service sampler. The barrels are interchangeable; they may be 1 m to 3 m long. The inside diameter is 7.6. cm and the wall is 1.6 mm thick. The barrel is made of high-strength stainless steel having a smooth interior surface. The nominal 7.6 cm inside diameter sampler samples unconsolidated deposits more accurately than samplers of small diameter (Hvorslev, 1948).

Gamma probe

Volume-weight of saturated sediment can be measured in place using a gamma probe that is properly calibrated in saturated material of similar specific gravity. This probe has a radioactive source (3 millicuries of radium 226) which emits gamma rays and a detector to pick up reflected radiation. The probe is connected electrically to a scaler by a coaxial cable (Figure 7.10.). The scaler readout is the total count for the time of measurement period (usually 1 minute). Extension pipes are threaded onto the gamma probe so that it can be shoved into sediment and removed. The 1.8 m aluminum extension pipes have a 38 mm outside diameter and a 6.35 mm wall thickness. The gamma probe has been used to penetrate as much as 4 m of sediment and to a total depth of 12 m. Additional extension pipes and coaxial cable would allow greater total depths. A yoke made from "vice-grip" clamps holds the extension pipe tightly and helps to remove it from the sediment (Figure 7.11.). The gamma probe, its accessory equipment and use and described in detail by McHenry (1962), Heinemann (1962), and McHenry and Dendy (1964).

A-frame and reel

An A-frame and a reel (US Geological Survey type B-50) are mounted on one of the boats for lowering and raising the gamma probe and sediment samplers (Figures 7.1. and 7.2.). The dial on the reel is used to indicate depth of the sampler. The A-frame is about 2.4 m tall and its top extends 0.3 m over the side of the boat. The hoist is used only when the two boats are fastened together by the two planks. A rack on the side of the A-frame holds the gamma probe extension pipes when they are not in use.

SEDIMENT VOLUME-WEIGHT DETERMINATIONS AND SAMPLES

Volume-weight determinations

Sediment volume-weight or dry bulk density can be determined by two methods; in place using a gamma density probe, or by removing an undisturbed sample of known volume and determining its dry weight. The gamma probe method is faster and more accurate but it can be used only in saturated sediment. Another disadvantage is that a few samples must

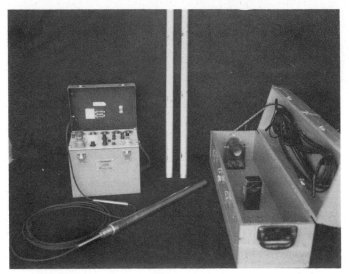

Fig 7.10. Scaler connected to gamma probe by coaxial cable extension pipe, and case for probe.

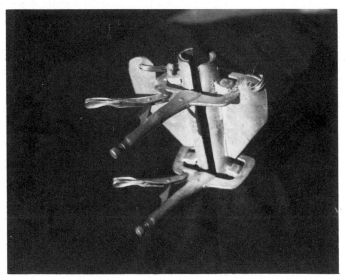

Fig 7.11. Yoke that clamps onto extension pipe. The cable on the yoke is attached to the hoist cable

still to be taken from each reservoir for particle size and specific gravity analyses.

Volume-weight measurements and sediment samples are tak at intervals along range lines. The number of samples and measurement locations depends on the accuracy desired and t

192

variability of the sediment. The entire depth of sediment deposit must be sampled and its volume-weight measured to accurately determine sediment accumulation, because consolidation occurs with time throughout the sediment depth. Gamma probe measurements are limited to the depth the probe can be shoved into the deposited sediment, usually 4 m or less. Readings generally are taken at 0.3 m depth intervals.

Sediment samples

Undisturbed samples of sediment deposits are taken by lowering the piston-type volumetric sampler (with the piston flush with lower end) vertically to the sediment surface. With the piston held at the sediment surface with its rope, the sampler barrel is driven into the sediment by repeatedly dropping the driving weight. When the barrel has penetrated the sediment or is full, the sampler is withdrawn from the sediment, using the reel and A-frame on the boat, making sure the piston does not move in relation to the sampler. With the sampler lying horizontally on the boat, relatively undisturbed samples can be obtained when the sediment is extruded from the barrel by pushing on the piston rod. The sampler is limited by the length of the barrel, usually 3 m or less. A 10 cm sample is taken from the core about every 0.3 m. Sample length and its depth in the deposit are determined by measuring the piston rod displacement. Samples are placed directly into plastic sample cartons, capped, and numbered. Sample numbers are recorded in a notebook and described as to reservoir, range, distance to range end, and depth. The rest of the core is observed closely for texture, air and water pockets, stratification, and organic matter. Such information along with total length of sediment in barrel, date, and daily weather conditions are also recorded in the notebook.
The sediment samples are usually analyzed in the laboratory for particle-size distribution, specific gravity, dry weight, and spectrographic data. Particle-size distribution may be determined by the hydrometer or pipette method before drying the samples. The specific gravity, measured using a pycnometer, is used to determine any shift in the calibration of the gamma probe. Spectrographic analyses may be performed on several samples to determine content of the elements adversely affecting absorption of gamma radiation from the gamma probe. These elements are iron, calcium, manganese, strontium and barium.

COMPUTATIONS

Capacity

After all the survey data have been gathered, a topographic map can be constructed. First, the intercepts of the contours and ranges are plotted. Contour lines are then drawn between ranges with the aid of other data such as point elevations of the sediment surface (deltaic deposits) and water elevation contour from the plane table sheet. This map is used for computing the present reservoir capacity. As mentioned earlier, there are several methods of computing reservoir capacity. Heinemann and Dvorak (1965) have shown that the stage-area curve is the most direct, simple, accurate, and uniformly adaptable method. The area inside each contour li is determined and plotted versus its elevation or stage. Th is done for each segment (that portion of the reservoir betw two or more ranges) and then for the total reservoir. In th stage-area curve method, the curve is integrated with a planimeter or an integrimeter with respect to stage. The instrument units are then converted to volume in cubic metre

Other computational methods have also been compared and evaluated by Heinemann and Dvorak (1965). Some are almost a good as the stage-area curve method. The average-contour ar method closely approximates the results of the stage-area curve method if small contour intervals are used and if the areas being averaged are not widely different. The average-contour area method uses a straight line approximation for integrating each increment of depth. This method is easily programmed for computer use to quickly calculate reservoir capacities, sediment volumes, and sediment distribution from stage-area data.

Sediment volume and sediment distribution

There are also several methods of computing sediment volumes. Sediment volume may be computed directly from cros sectional areas of sediment by the methods described earlier The most accurate method for determining sediment volume, however, is to take the difference between the original capacity and any subsequent capacity. This can be done for each increment of elevation in each segment of the reservoir The sum of all sediment volume increments is the total sediment volume. The increments of sediment volume can be summed horizontally and vertically to give vertical and horizontal sediment distributions, respectively. The distribution of sediment is usually expressed as a percentag of the total sediment volume for a percentage of original depth (vertical distribution) or for a percentage of total distance from the dam (horizontal distribution). The sediment volume is also used with the sediment volume-weight to determine the true rate of sediment accumulation by weigh

Volume weight

If gamma probe readings are taken, they may be converted to wet densities using the calibration curve for the particular gamma probe and scaler involved. Volume-weight o a dry weight basis is then computed using the following formula:

$$\gamma_d = \frac{G_s(\gamma_w - 1.0)}{G_s - 1}$$

where γ_d is dry volume-weight in g/cm^3

γ_w is wet density in g/cm^3

G_s is specific gravity of sample at same location and depth; and

1.0 = weight of water in g/cm^3 at 0°C; use 62.4 for lbs/ft^3.

Volume-weight of the undisturbed samples can be determined by drying the sample or known portion thereof and the dividing the dry weight by the known sample volume.

Sediment weight

Sediment accumulation values should be reported on weight rather than a volume basis because the volume-weight of sediment varies within a reservoir and between reservoirs. The sediment also consolidates with time and with depth below the sediment surface. The increase in total weight of the reservoir sediment with time reflects the true rate of accumulation and, when adjusted for reservoir trap efficiency, is the sediment yield from the watershed.

The weight of sediment in a reservoir is computed from sediment volumes and weighted averages of volume-weights. A weighted average is computed for each range, based on the cross-sectional area that each volume-weight reading represents. The weighted average volume-weight for each segment is then based on the weighted average volume-weight of each bounding range, which is weighted according to the cross-sectional area of each range represented in the segment. The weighted average volume-weight for each segment times the sediment volume in each segment equals the weight of sediment in each segment. The total weight of sediment in the reservoir is then the sum of all segment weights. Figure 7.12. and Table 7.1. show an example of these computations for a segment of the reservoir bounded by two ranges.

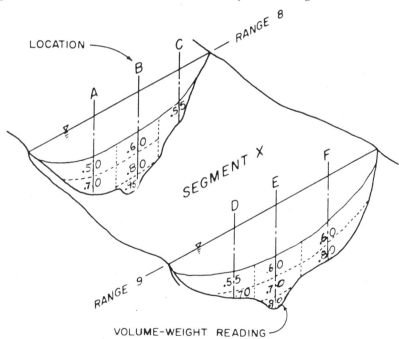

Fig 7.12. Possible volume-weights in a typical segment X, bounded by ranges 8 and 9.

Table 7.1. Example of weighted average volume-weight computations for one segment bounded by two ranges

Range Number	Location on range	Volume-weight Reading g/cm^3	Representative X-sectional area m^2	Product of Column 3 and Column 4
8	A	.50	20	10
		.70	10	7
	B	.60	20	12
		.80	20	16
		.95	10	9.5
	C	.55	15	8.25
			95	62.75
9	D	.55	20	11
		.70	5	3.5
	E	.60	20	12
		.70	10	7
		.90	5	4.5
	F	.60	20	12
		.80	10	8
			90	58

Weighted averages:

    Range 8:   62.75 ÷ 95 = 0.660 g/cm^3
    Range 9:   58.00 ÷ 90 = 0.644 g/cm^3
    Segment (sum of Ranges 1 and 2): 120.75 ÷ 185 = 0.653 g/cm^3

Sediment weight for segment = sediment volume x 0.653 g/cm^3 or 653 kg/m^3.

Sediment yield

        In order to determine the average annual sediment yield
from the contributing watershed, the weight of deposited
sediment must be adjusted for reservoir sediment trap
efficiency. This adjustment is needed because not all of the
incoming sediment is trapped and deposited in the reservoir,
usually some is passed through the spillway. By definition,
trap efficiency is the portion of the incoming sediment
(sediment yield) that is deposited or trapped in the
reservoir, usually expressed in percent. Sediment yield,
therefore, on a weight basis is calculated as:

$$\text{Sediment yield} = \frac{\text{Weight of Deposited Sediment x 100}}{\text{Trap Efficiency}}$$

The trap-efficiency value can usually be determined quite
easily from Brune's (1953) or Heinmann's (1981) trap
efficiency curves (Figure 7.13.). This requires knowledge of
the average reservoir capacity up to the emergency spillway

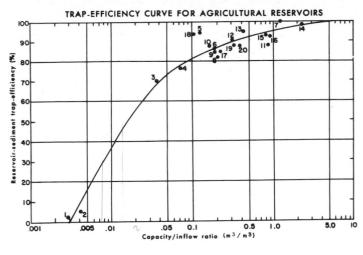

TRAP-EFFICIENCY CURVE FOR AGRICULTURAL RESERVOIRS

1. Halbert Rock Res.#1, Tex.   11. Upper Hocking #1, Ohio
2. Halbert Rock Res.#3, Tex.   12. Plum Ck. #4, Ky.
3. Halbert Earth Res.#1, Tex.   13. Six Mile Ck. #6, Ark.
4. Lexington, N.C.   14. Escondido #1, Tex.
5. Isaqueena, S.C.   15. Double Ck. #5, Okla.
6. T & P Reservoir, Tex.   16. Brownell #1, Neb.
7. H. Lage, Iowa   17. Highland Ck., Cal.
8. Third Ck. #7A, N.C.   18. Ashland, Mo.
9. N. Fork Broad R., #14, Ga.   19. Callahan C-1, Mo.
10.Salem Fork #11A, W. VA.   20. Bailey, Mo.

Figure 7.13.   Curve for predicting agricultural reservoir-
sediment trap efficiency (from Heinemann, 1978).

level for the period included in the determination of the
weight of deposited sediment and of the average annual inflow.
The ratio of these two values, capacity/inflow, can then be
used to obtain the trap efficiency value from the curves. The
average annual sediment yield is the sediment yield thus
calculated divided by the age of the reservoir and the
watershed area. A reservoir survey will be more useful if a
complete discription of watershed conditions that affected
sediment yield are included.

SUMMARY

    In summary, the general procedure for making reservoir
sedimentation surveys has been described and the following
items are emphasized.

1. Accuracy problems associated with many past reservoir
    sedimentation surveys can be avoided by systematic
    procedures and techniques outlined in this chapter.

2. Each reservoir must be considered individually and its
    base line and range layout carefully planned before
    the survey begins.

3. A plane table and alidade can be used to map readily and accurately the exposed deltaic deposits, spillway and shoreline contours and range ends. Use of an engineer's level is suggested for vertical control.

4. A sonic depth recorder electrically pulsed by a switch on the cable meter can be used to automatically plot water depth versus distance as it is moved across the reservoir on each range. The transducer for the depth recorder is mounted so that water depths as shallow as 0.5 m can be measured. Optional manual sounding techniques may be used instead.

5. A gamma probe can be used to measure the volume-weight of the saturated deposited sediment in place. A piston type sampler can be used to also remove an undisturbed sample for further determination of its physical properties.

6. Reservoir capacity is best determined by the stage-area curve method - a direct, simple, and accurate method that is easily adapted to all reservoirs. Alternately, the average-contour method, which closely approximates the results of the stage-area curve method, can be easily adapted to computer use to determine capacities and sediment volumes and distribution.

7. Sediment volume is computed as the difference between the original reservoir capacity and a subsequent capacity. The difference in volume between any two subsequent surveyed capacities will not accurately estimate additional sediment accumulation because earlier deposits continue to consolidate.

8. Sediment accumulation should be expressed as a total weight of sediment. The total weight of sediment is computed from the sediment volume and its weighted average volume-weight.

9. Sediment yield can be determined from the deposited sediment weight by applying a reservoir sediment trap efficiency value to account for the sediment that passed through the reservoir.

REFERENCES

Brune, G.M., 1953, Trap efficiency of reservoirs, *Transactions, American Geophysical Union*, 34, 407-418.

Eakin, H.M., 1939, Silting of reservoirs, *US Department of Agriculture Technical Bulletin* 524.

Heinemann, H.G., 1981, A new sediment trap efficiency curve for small reservoirs. *Water Resources Bulletin*, 17, 825-30.

Heinemann, H.G., 1962, Using the gamma probe to determine the volume-weight of reservoir sediment, *International Association of Scientific Hydrology Publication* No. 59, 410-423.

Heinemann, H.G., and Dvorak, V.I., 1965, Improved volumetric survey and computation procedures for small reservoirs, *Proceedings of the Federal Inter-Agency Sedimentation Conference, 1963,* Miscellanous Publication 970, Agricultural Research Service, US Department of Agriculture, Washington, D.C., 845-856.

Heinemann, H.G., and Rausch, D.L., 1971, Discussion of chapter III, Sediment measurement techniques: reservoir deposits, by J. Lara, *Proceedings, American Society of Civil Engineers, Journal of the Hydraulics Division*, 97, 1551-1561.

Hvorslev, M.J., 1948, Subsurface exploration and sampling of soils for civil engineering purposes, *US Army Corps of Engineers, Research Report*.

McHenry, J.R., 1962, Determination of densities of reservoir sediments in situ with a gamma probe, *US Department of Agriculture Report* ARS 41-53.

McHenry, J.R., and Dendy, F.E., 1964, Measurement of sediment density by attenuation of transmitted gamma rays. *Soil Science Society of America Proceedings*, 28, 817-822.

McHenry, J.R., Hawks, P.H., Harmon, W.C., Kelly, W.J., Gill, A.C., and Heinemann, H.G., 1971, Determination of sediment density with a gamma probe: a manual of theory, operation, and maintenance for technical operations model 497, *US Department of Agriculture Report* ARS 41-183.

Rausch, D.L., and Heinemann, H.G., 1968, Reservoir sedimentation survey methods, *University of Missouri Agricultural Experimental Station, Research Bulletin* 939.

Rausch, D.L., and Heinemann, H.G., 1976, Reservoir sedimentation survey methods in *Hydrological techniques for upstream conservation, FAO Conservation Guide* No.2, 29-43.

Timblin, L.O., and Florey, Q.L., 1957, Density measurement of saturated submersed sediment by gamma ray scattering, *Chemical Engineering Laboratory Report* No. SI-11.

US Department of Agriculture, Soil Conservation Service, 1973, *National Engineering Handbook*, Section 3 (Sedimentation), Chapter 7, Field Investigations and Surveys.

Vanoni, V.A. (ed), 1975, *Sedimentation Engineering, American Society of Civil Engineers Manuals and Reports on Engineering Practice* No. 54.

8.

Reservoir trap efficiency

H. G. Heinemann

Introduction

Even after hundreds of years of designing and
constructing dams and reservoirs, man does not completely
understand the sedimentation processes in reservoirs. For
example, we still need to know more about why or how the
sediment is deposited where it is. We also need to improve
our accuracy in estimating the long-term sediment trap
efficiency for proposed small reservoirs under a variety of
environmental conditions. This improved technology is
necessary because good reservoir sites are scarce and
constitute a valuable natural resource that must be protected
and used wisely.
Because of limited sites and increasing construction
costs, we must carefully design and build each reservoir to
best accomplish its specific objectives - for soil and water
conservation, irrigation, domestic or animal watering, fish
farming, recreation, or protecting and enhancing our
environment. To optimize the effectiveness of each
reservoir, we must be able to predict the rate of reservoir
sedimentation processes, especially reservoir-sediment trap
efficiency. Reservoir-sediment trap efficiency is the
fraction of the sediment transported into a reservoir that
is deposited in that reservoir, usually expressed as a
percentage. Knowledge of this process is needed to control
the sediment accumulation and thereby the life of the
reservoir, and to assure its proper operation.
This paper contains an explanation of what happens in
agricultural reservoirs (most are from 3 to 980 $m^3.10^4$ in
capacity) during an inflow event, and using a flow diagram
(Figure 8.1), the various parameters that influence
sediment trap efficiency are discussed. Included is a

Contribution of the Agricultural Research Service, USDA.

Research Hydraulic Engineer, Hydrograph Laboratory,
Beltsville, Maryland, USA (deceased).

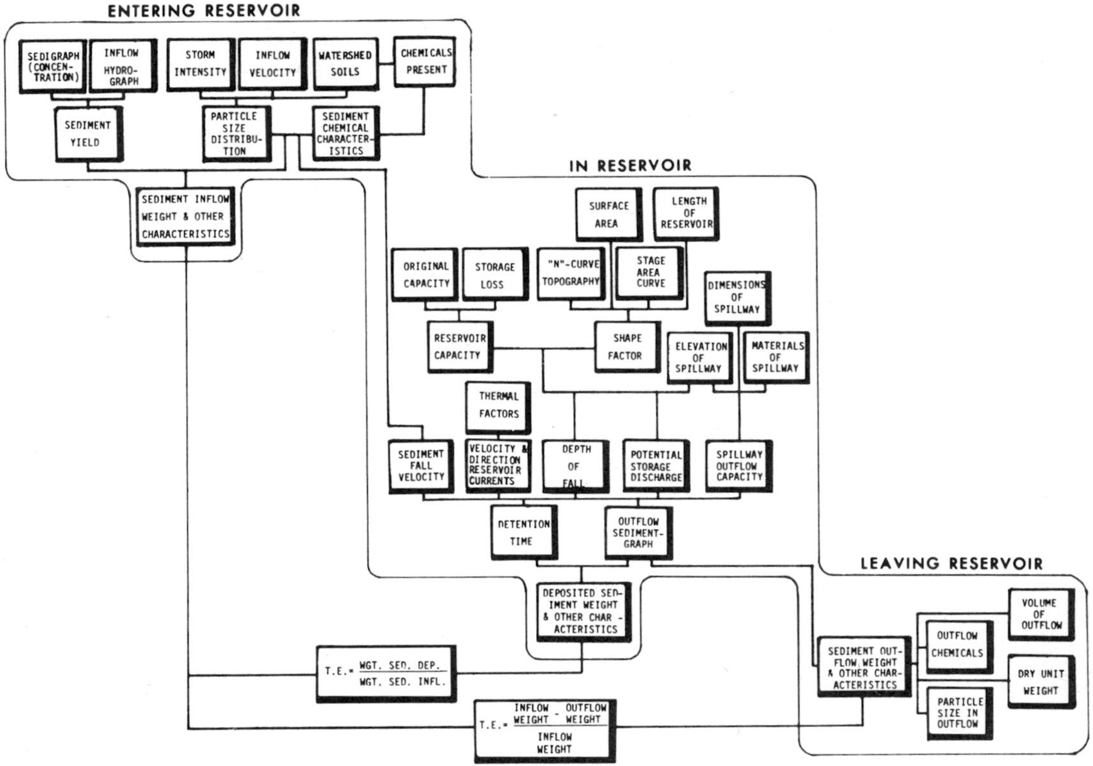

Fig 8.1. Reservoir-sediment trap-efficiency parameter flow chart.

literature review of publications that have helped advance
the state of the art to our current level of knowledge.
The individual reports can then be compared with the flow
diagram to elevate their completeness and adequacy in
estimating the sediment trap efficiency in a proposed
reservoir.

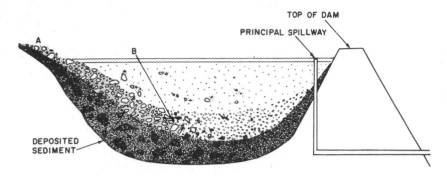

Figure 8.2.   Depicting storm runoff moving into and in a
reservoir.

Reservoir Sedimentation

Sediment movement in reservoirs

     When storm runoff enters a reservoir (Figure 8.2,
point A), the inflow is spread over a larger channel/
reservoir cross-section and its velocity is quickly reduced.
This reduces the transport energy and causes the large
sediment particles and aggregates to settle to the bottom.
The remainder of the inflow moves along the bottom of the
reservoir towards the dam until it reaches an elevation in
the reservoir where the density of the inflow equals the
density of the reservoir water. As the inflow velocity is
further reduced, the larger particles left in the remaining
flow will settle to the bottom, decreasing the inflow's
density. Some of the flow may move horizontally into the
reservoir before the bulk of the remaining flow. (This is
a very dynamic process that is constantly changing and
adjusting). When the flow reaches this point of equal
density (point B), it flows horizontally into the reservoir
(somewhat like a wedge) between the lighter and denser water,
and raises the water in the reservoir above it. In a full
reservoir equipped with a surface discharge principal
spillway, the upper level of water (the highest quality
water in the reservoir) would be discharged.
     The density of the storm inflow depends on its
temperature and sediment concentration. The sediment
concentration is often the more important parameter in
reservoirs, because the temperature difference between
storm inflow and reservoir water is usually not large. For
example, only 1000 ppm of sediment is needed to equal the
density difference caused by the reservoir water being $5.5^{\circ}C$
cooler than the inflow (in the 10° to $27^{\circ}C$ range).

Trap efficiency

Good estimates of sediment trap efficiency of proposed reservoirs are important because the volume of sediment trapped during the design life of the structure must be provided for in the reservoir capacity; this plus water storage for the design storm are the two components that govern the ultimate size of the reservoir. If the trap-efficiency estimate is less than it should be, the reservoir is underdesigned and its capacity will be filled with sediment too soon and its useful life will be shortene If the estimate is larger than necessary, the reservoir is overdesigned and money will be wasted constructing too larg a structure, and the reservoir may not function at an optimal level. Furthermore, we must learn the controlling physical dimensions or characteristics for reservoir-sedime trap efficiency and how to better change these parameters. We can then incorporate the proper controls into the design of each reservoir so that it will trap the percentage of incoming sediment needed to accomplish the primary objectiv If the primary objective of the reservoir is for domestic water supply, irrigation, emergency water for fighting fire recreation, fish farming of certain species, etc., the reservoir designer would want to limit the amount of sedime trapped.

To determine reservoir-sediment trap efficiency of existing reservoirs requires an accurate measurement of all sediment transported into the reservoir as well as the sediment discharged through the spillways. This requires flow measurements and samples. An as alternative to flow measurements and samples of the inflow and outflow, we can measure only one and determining the sediment retained by making good reservoir sedimentation surveys of the deposite sediment volume and its volume-weight.

Reservoir-sediment trap efficiency is best discussed by considering the parameters in their respective zone of influence. In sequence these are a characterization of: (L) inflowing watershed runoff and sediment, (2) reservoir storage dimensions and properties, and (3) discharge locatio and capabilities. Using the parameter flowchart (Figure 8. as an aid, we can better follow this sequence.

The storm runoff from the contributing watershed will flow into a reservoir at a variable rate for the water component and a different variable rate for the sediment concentration component. The hydrograph will show the wate inflow as a function of time, and the sediment graph will show the sediment concentration also as a function of time. When used together, we can compute the sediment yield to the reservoir for the storm or a unit of time. The storm intensity and inflow velocity control the size of sediment particles eroded on the watershed uplands and channels and transported to the reservoir. Of course, certain chemicals in the soil or water may cause flocculation or aggregation and affect the particle or aggregate size, density and fall velocity. All of these characterize the inflow and determines the amount of sediment moving into the reservoir. In the reservoir, the capacity and its configuration are very important parameters. We do not know which reservoir

configuration parameter will be most important. These might
take on different degrees of importance depending on the
size of the reservoir. For example, in a very small reservoir,
the sediment inflow will be close to the dam and there will
be little opportunity for even the larger particles to be
deposited far from the dam. The situation will be quite
different in a large reservoir. Another important factor
governing reservoir dynamics is thermal stratification.
The spillway characteristics of elevation, size, design and
roughness will control the spillway outflow capacity. This,
with sediment fall velocity, depth of fall, storage to be
discharged, temperature, and current velocity will govern
detention time and the outflow sediment graph. They, in
turn, will control the amount of sediment that will be
deposited and the residual - the amount that will be
released from the reservoir.

The spillway location, elevation, and capacity will
greatly influence the sediment outflow. Usually, the
sediment passing through a reservoir will be clays and
highly dispersed particles. The sediment discharge or
outflow can be characterized as to volume, particle-size
distribution, adsorbed chemicals, and dry volume-weight.

Trap/efficiency can be determined in several ways.
Reservoir-sediment trap efficiency (E) (usually expressed
in percent) is the ratio of the weight of sediment (S)
coming into a reservoir to the weight that is trapped
therein

$$E = \frac{S\ retained}{S\ Yield} \quad or \quad E = \frac{S\ inflow - S\ outflow}{S\ inflow}$$

Sediment yield and sediment inflow are the same parameter.

Evolution of current state of the art

Through the early years, the methods for estimating
reservoir-sediment trap efficiency remained relatively
unchanged. They were based primarily on empirical
relationships.

Hazen (1904) has been credited for developing the first
real theory on the operation of sedimentation basins. This
was a further development of some ideas proposed by Seddon
in 1889. Hazen developed his concepts by considering a
series of increasingly complex hydraulic situations and
assumptions. His fundamental proposition was that a
particle of sediment settles at a velocity that depends
upon its size and weight, and upon the viscosity of the
water. Second in importance was the density of sediment in
the water immediately above the bottom.

Hazen (1914) first introduced reservoir storage, or
capacity, in terms of runoff per square mile of tributary
area - the C/1 ratio. However, he used this term in
connection with reservoir storage requirements, instead of
reservoir-sediment trap efficiency.

Brune and Allen (1941) reported a good relationship
(Figure 8.3) between the percentage of eroded soil (gross
erosion) caught in the reservoir and a capacity-watershed
ratio, expressed as storage capacity per square mile of
drainage area. They used 23 reservoirs from Texas to

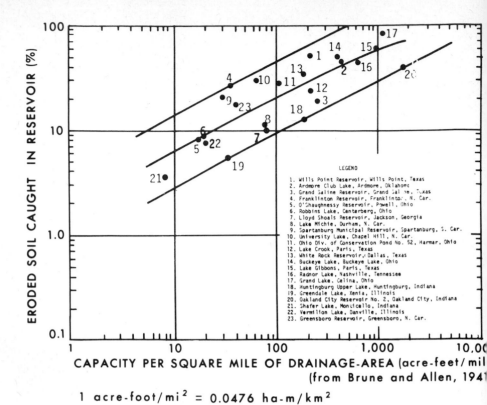

LEGEND

1. Wills Point Reservoir, Wills Point, Texas
2. Ardmore Club Lake, Ardmore, Oklahoma
3. Grand Saline Reservoir, Grand Saline, Texas
4. Franklinton Reservoir, Franklinton, N. Car.
5. O'Shaughnessy Reservoir, Powell, Ohio
6. Robbins Lake, Centerberg, Ohio
7. Lloyd Shoals Reservoir, Jackson, Georgia
8. Lake Michie, Durham, N. Car.
9. Spartanburg Municipal Reservoir, Spartanburg, S. Car.
10. University Lake, Chapel Hill, N. Car.
11. Ohio Div. of Conservation Pond No. 52, Harmar, Ohio
12. Lake Crook, Paris, Texas
13. White Rock Reservoir, Dallas, Texas
14. Buckeye Lake, Buckeye Lake, Ohio
15. Lake Gibbons, Paris, Texas
16. Radnor Lake, Nashville, Tennessee
17. Grand Lake, Celina, Ohio
18. Huntingburg Upper Lake, Huntingburg, Indiana
19. Greendale Lake, Xenia, Illinois
20. Oakland City Reservoir No. 2, Oakland City, Indiana
21. Shafer Lake, Monticello, Indiana
22. Vermilion Lake, Danville, Illinois
23. Greensboro Reservoir, Greensboro, N. Car.

CAPACITY PER SQUARE MILE OF DRAINAGE-AREA (acre-feet/mi
(from Brune and Allen, 1941

1 acre-foot/mi^2 = 0.0476 ha-m/km^2

Fig 8.3. Percentage of eroded soil caught in reservoir as function of reservoir capacity-drainage-area ratios. (from Brune and Allen, 1941).

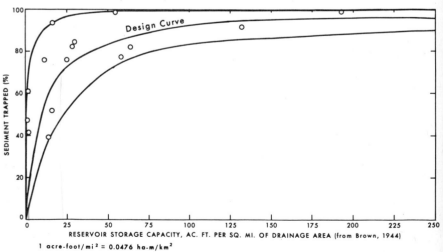

1 acre-foot/mi^2 = 0.0476 ha-m/km^2

Fig 8.4. Relation of reservoir trap efficiency to reservoir storage capacity per sq. mi. of drainage area. (from Brown, 1944).

206

Ohio as a basis for their work, one of the first reservoir-sediment trap efficiency studies.

Brown (1943) reported that "Study of reservoir silting, both in this country and abroad, has shown that one of the most important factors governing the annual rate of storage loss is the ratio between the original storage capacity of the reservoir and the inflow of water from the drainage basin". Brown then separated his data into groups depending upon the original storage of the reservoir per square mile of drainage area. Those with the lowest capacity per unit drainage area had the highest rate of storage loss due to sediment deposition.

Brown (1944) developed a curve (Figure 8.4.) showing the relationship between reservoir-sediment trap efficiency, and the ratio of capacity (original) to watershed drainage area. His curve was based on data from 15 reservoirs. Brown enclosed his data spread in an envelope of curves and attributed the higher percentage trap efficiency curve to smaller and more variable runoff, coarse, or highly coagulated sediments, and structures with greater storage capacity.

Campe (1945) developed several theories regarding the settling velocities of particles (based on Stokes' law) in an idealized, rectangular, continuous flow basin. His studies included work on the particle drag coefficients; hinderance of settling due to close proximity of other particles; factors influencing settling velocity; effect of flocculation on settling velocity; settling path; resuspension; and effects of turbulence, water depth, and detention period on deposition. Camp also developed a family of trap efficiency curves based on settling velocity, rate of outflow from the basin, and surface area of the basin.

Churchill (1948) outlined the method used by the Tennessee Valley Authority in estimating reservoir-sediment trap efficiency. He used a ratio of period of retention to transmission velocity as the Sedimentation Index and related it to trap efficiency. Churchill used two sizeable reservoirs, Hales Bar and Wilson, as the principal basis for his curves (Figure 8.5.). These curves fitted the relatively fine-grained sediment found in the Tennessee Valley; however, different particle sizes will result in different relationships. The components of the sedimentation index, period of retention, and mean velocity of flow through the reservoir, are not generally available for most reservoirs.

Brune (1953) used data from 40 reservoirs (44 periods of time) to develop trap efficiency curves (Figure 8.6.) based on the capacity-inflow ratio. Brune originally constructed envelope curves for normally ponded reservoirs, but these were named to reflect the expected character of the sediment, such as highly flocculated and coarse sediment versus very fine sediment. Brune's curves have been used more widely than other methods, especially for estimating the trap efficiency of small reservoirs. Brune concluded that the capacity-inflow (C/1) ratio is much better than the capacity-watershed (C/W) ratio formerly used.

Guy et. al. (1958) described the plan of operations and

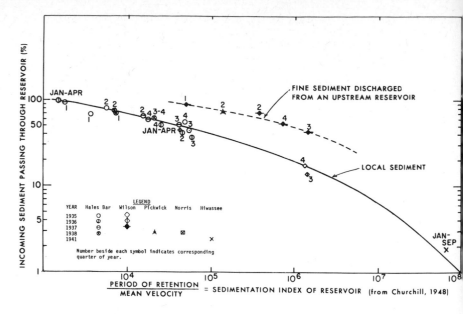

Fig 8.5. Relation of reservoir sedimentation index to percent of incoming sediment passing through reservoir. (from Churchill, 1948).

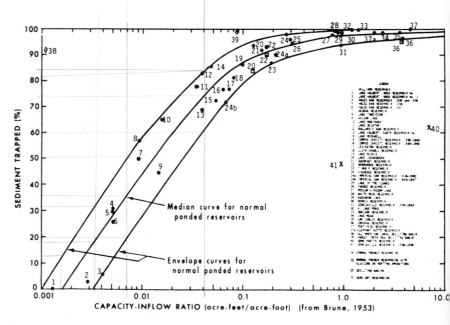

Fig 8.6. Trap efficiency as a function of capacity-inflow ratio, type of reservoir, and method of operation. (from Brune, 1953).

some of the details of a cooperative reservoir-sediment trap efficiency study financed primarily by the Soil Conservation Service, with participation also by the US Geological Survey and the Agricultural Research Service. The authors discussed how the sediment trap efficiency depends on settling velocity of sediment and retention time in the reservoir. They included information on 12 reservoirs in 11 states and trap efficiency estimates of 10 of these reservoirs (to June 30, 1957).

Heinemann and Reynolds (1962) reported on the same cooperative reservoir-sediment trap efficiency study and listed 26 basic measurements or parameters that might influence trap efficiency, including a characterization of the inflow, the reservoir itself, and the outflow structure. They used a form of sedimentation information curve to study reservoir sedimentation and the effect of the size of principal spillways on the trap efficiency and sediment deposition of several small reservoirs.

Gottschalk (1965) more fully explained the above mentioned cooperative study and the use of such data in designing small floodwater retarding structures. He also showed the measured trap efficiencies for 18 small reservoirs - these data points fell between or below Brune's envelope curves, indicating a possible overestimation of trap efficiency.

Beer, Farnham, and Heinemann (1966) evaluated sedimentation prediction techniques in western Ohio using data from a detailed study of 24 small reservoirs and their watersheds. Their results suggested that capacity-inflow may not be the best estimator of trap efficiency for reservoirs in the loess area. A regression correlation showed that a reservoir capacity-watershed area term was about twice as good as an indicator of trap efficiency as capacity-inflow in the loess area.

Borland (1971) used the basic Churchill (1948) curve and added 15 data points representing desilting basins and semidry reservoirs. He concluded that this relationship was more applicable than Brune's curves for estimating trap efficiencies for desilting and semidry reservoirs.

Dendy (1974) summarized the results from the cooperative study by the Soil Conservation Service, US Geological Survey, and the Agricultural Research Service referenced earlier. These studies were conducted on 11 normally ponded and six dry reservoirs in the southern United States. Dendy makes the point that trap efficiency usually depends on the reservoir's ability to trap the silt-size and smaller sediments. He also emphasized that all but one data point for these reservoirs plotted below Brune's (1953) curve (Figure 8.6.).

Chen (1975), in addition to providing a good general review of the state of the art of trap efficiency, developed a series of curves (Figure 8.7.) for various particle sizes (d) showing trap efficiency related to the ratio of basin area to outflow rate (A/Qo). These showed that clay-size particles require excessively large basin dimensions to be trapped, unless chemical flocculants are added to increase settling velocity. He also compared Brune's curves (1953) and Churchill's curve (1948) with the trap efficiency

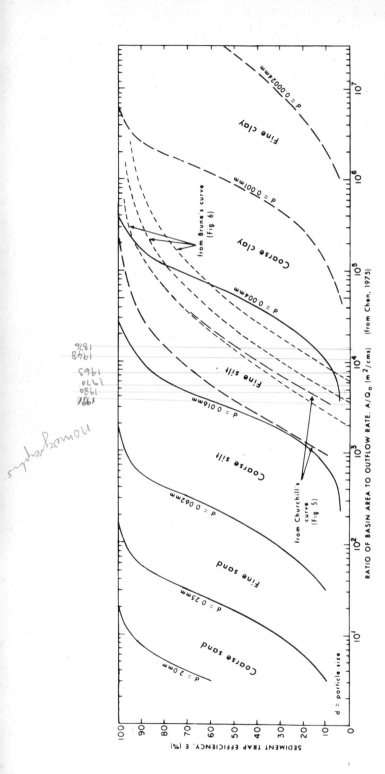

Fig 8.7. Trap efficiency as a function of ratio of basin area to outflow rate. (from Chen, 1975).

curves developed by Camp (1945) and found that they were
compatible in the silt range. He concluded that for a given
basin dimension, both Brune's and Churchill's curves tend
to underestimate trap efficiency for coarser material, but
overestimate it for finer sediments. He also concluded that
trap efficiency increases as the basin outflow rate
decreases and that outflow rate is governed by basin
storage capacity and the configuration and capacity of
spillways and release outlets.

Bondurant, Brockway and Brown (1975) reported trap
efficiency information on two irrigation return flow ponds.
They found that sediment removal efficiency correlated well
with flow rate and sediment concentration. They also showed
the sediment particle size distribution of one pond.

Rausch and Heinemann (1975) reported on their trap-
efficiency studies of three reservoirs in central Missouri,
the first study of its kind on a storm basis. Their study
yielded 48 data points for a regression analysis which
showed that the most important parameters were reservoir
detention time and particle size of the inflowing sediment.
Peak inflow rate was substituted for sediment particle size
since they found a high direct correlation between these
two parameters. Storm runoff volume, sediment yield,
reservoir capacity and drainage area also improved the
prediction of trap efficiency.

Pennell and Larson (1976) developed a mathematical model
to evaluate reservoir design factors and their significance
and effects on trap efficiency. They showed that the most
significant design factors are capacity, basin depth, and
length of detention time.

Curtis and McCuen (1977) developed a model, based on
Camp's (1945) approach, which shows the effect of four
parameters on reservoir-sediment trap efficiency:

(1) particle size distribution: trap efficiencies are
     higher in reservoirs below watersheds with eroded
     soil composed of a high portion of large, heavy
     particles.

(2) initial basin storage: the more runoff already
     stored, the less available for additional runoff
     and, therefore, the lower the trap efficiency.

(3). outflow: the larger the outflow, the lower the
     trap efficiency.

(4) basin depth: when the volume of water stored is
     held constant, the shallower depths gave higher
     trap efficiency.

Their model was developed on the basis of small idealized
settling tanks and they found no data that could be used for
verification or calibration.

Ward, Haan, and Barfield (1977a) conducted an extensive
literature review on the sedimentation processes in detention
basins and developed a mathematical model describing the
sedimentation characteristics of such small basins. This
model is very comprehensive and uses as basic input the
inflow hydrograph, inflow sediment graph, sediment particle
size distribution, detention basin stage-area relationship,

and detention basin stage-discharge relationship. The model
is used to route the water-sediment mixture through the basin
and in the process, estimates the outflow sediment
concentration, sediment distribution, and the sediment trap
efficiency.
With respect to sediment particle size, the percent finer
than 0.02 mm was the most critical in determining the
performance of a sediment basin.
    Ward, Haan, and Barfield (1977b) evaluated the most
commonly used trap efficiency methods, emphasizing that
most are empirical. The authors further explained their
DEPOSITS model - a mathematical simulation model for
predicting the sediment processes occurring in small
reservoirs. They also stressed the importance of
aggregation and flocculation in settling of particles, and
the need for suitable field data for testing theory and
models.
    Ward, Haan, and Barfield (1977c) reported additional
studies with their DEPOSITS model and limiting conditions
for its use. They used their model to develop regression
equations for estimating reservoir-sediment trap efficiency
for different kinds of small basins, especially those used
to control sediment from strip mines and urban areas.
    Schiebe and Dendy (1978) used a small laboratory
reservoir to study residence or detention time under
several different inflow and reservoir stratification
conditions in an effort to learn how better to control
detention time in different kinds of reservoirs. They also
verified that the time available for sediment settling can
be changed by manipulating the location and operation of
the reservoir outlet.
    Of the above studies on reservoir-sediment trap efficiency
only five authors; Chen (1975), Rausch and Heinemann (1975),
Curtis and McCuen (1977), Ward, Haan, and Barfield (1977a, b,
and c), and Schiebe and Dendy (1978), considered trap
efficiency in its entire context - that is, considered and
characterized the inflow, the reservoir storage dimensions
and its effect, and the outflow. Some of these are field
studies and others are primarily theoretical studies, and
the mix is a healthy one which should lead to more progress.
Actual verification is still needed for the theoretical model.

Application of trap efficiency in design

    As discussed earlier, the estimated volume of sediment
that will be trapped in a reservoir is one of the two
components determining the design capacity of the reservoir.
This estimate of trapped sediment is made by multiplying
estimated reservoir-sediment trap efficiency values times the
sediment yield to the reservoir site for the design life of
the structure.
    The estimated reservoir-sediment trap efficiency value
can be determined by any of the methods just described. The
method selected will probably depend on the users experience
with these methods and the availability of data. The method
that has been used more than any other is Brune's (1953)
curves. In this method, trap efficiency is estimated on the
basis of the ratio of reservoir average capacity to the
average annual inflow using the following procedure.

A   Estimate the total required capacity of the reservoir
    for water and sediment storage (Roehl, 1975). Since an
    actual value for the total capacity cannot be obtained
    until final design is completed, and approximation of
    the total capacity is made as follows.

    1. Estimate the sediment yield to the reservoir site,
        using procedures outlined elsewhere in this manual,
        the ASCE Sedimentation Manual (ASCE, 1975), or the
        USDA publication, "Present and Prospective Technology
        for Predicting Sediment Yields and Sources" (UDSA,
        1975). If the reservoir objectives and design are
        to trap most of the sediment, multiply the sediment
        yield value times a large trap efficiency, but if
        the objectives and design are to trap a small
        percentage of the sediment inflow, multiply the
        sediment yield times a low trap efficiency value.
        This gives the required sediment storage for a
        short period or the design life, depending on the
        time span considered. Another alternative is to
        assume a reasonable and realistic volume of sediment
        storage that might be required for the design life
        of the structure. For example, 4 cm (from the entire
        watershed).

    2. Obtain an estimate of the required water detention
        storage of the design storm. For example, 12 cm.

    3. The sum of 1 and 2 is the estimated total original
        capacity of the reservoir. That is, 4 + 12 = 16 cm.

    4. Repeat step 1 (above) progressively by time increments,
        or for the entire design life in one calculation, to
        obtain a final capacity of the reservoir. This
        decrease in reservoir capacity must depend on the
        trapping of sediment in the reservoir during the time
        period (or periods) being considered. For simplicity
        here assume that all of the sediment storage
        allocation has been filled. The capacity of the
        reservoir at the end of the reservoir design life
        would then be, 0 + 12 = 12 cm.

B   Determing the average annual runoff into the reservoir,
    in the same units as above. This value may be obtained
    from the hydrologic analysis of the watershed or other
    available information. For purposes of this illustration,
    it is determined to be 40 cm (from the entire watershed).

C   Divide the approximate average total capacity, item A-3
    plus A-4 divided by 2, by the average annual runoff, item
    B above, to obtain the capacity-inflow (C/1) ratio. That
    is $\frac{16 + 12}{2}$ ÷ 40 = 0.350 = C/1 ratio.

D   The trap efficiency for a given C/1 ratio is determined
    on the vertical axis of Brune's curves (Figure 8.6.).
    The texture of the sediment should be estimated on the
    basis of the character of the watershed soils and the
    principal sources of sediment. Where incoming sediment
    is assumed to have a predominance of bed load or coarse
    material or is highly flocculated, the upper of Brune's
    curves should be used to determine the trap efficiency.
    If the incoming sediment is composed primarily of
    colloids, dispersed clays and fine silts, the lower
    curve should be used. The median curve is representative
    of incoming sediment consisting of a wide distribution
    of various grain sizes. This trap efficiency value then
    is the first approximation used in the preliminary
    designs. As the basic design values become established
    the above procedure is repeated and a refined estimated
    trap efficiency value is developed and used.

Research needs

    As indicated, there have been a number of theoretical
studies and data analyses pertaining to reservoir-sediment
trap efficiency. In general terms we know about the
reservoir sedimentation processes and trap efficiency, but
we lack specific quantitative data. We would be hard pressed
to design a reservoir that was to trap only 50% of the
incoming sediment, or to trap only the sediment larger than
a specific particle size on a given watershed. We need to
improve our understanding of the depositional process and
to improve prediction and control of sediment deposition in
reservoirs. We also need to learn how to better control and
predict reservoir-sediment trap efficiency.
    I know of only 20 small reservoirs in the USA that have
been studied and measured in sufficient detail to provide
usable trap efficiency data. These are:

    7 reservoirs from Brune's (1953) report. These have
    drainage areas less than 38.85 km^2 (15 mi^2), which is
    the limit of the ponded reservoirs in Dendy's (1974)
    report.

    10 reservoirs (ponded) from Dendy's (1974) report.
    This does not include Brownell No. 1-A, which is almost
    filled and functions more like a dry reservoir.

    3 reservoirs from Rausch and Heinemann's (1975) report.

    Some other data exist, but these reservoirs were sampled
and runoff measured during only a part of some storms.
Questions have been raised regarding the adequacy of those
measurements.
    Obviously, this lack of good usable data is a very serious
research deficiency in the USA, and this problem is being
addressed by conducting additional studies at Oxford,
Mississippi, and at Columbia, Missouri. Other studies have
been started by Dr. Haan at the University of Kentucky.

More information is also needed on sedimentation processes in small reservoirs between runoff events. There are many unanswered questions about sediment movement, resuspension, temperature effects, and changes with time. These items, too, are included in the studies at Oxford and Columbia. The effects of temperature on small reservoir sediment deposition is not understood. How important is it? What is the effect of temperature on density currents and can they be utilized to control reservoir-sediment trap efficiency?

The entire area of flocculation and aggregation needs to be studied with regard to trap efficiency and predicting its effect in proposed reservoirs. Can chemical flocculants and flow velocity controls be used efficiently and practically to induce deposition where it is desired? Changes in trap efficiency and compaction of sediment with time must be investigated further. Further studies should be based on sediment weight. Sediment volumes alone are not very helpful because they change, depending on the degree of compaction experienced. For this reason, volume units are sometimes misleading.

Our future studies should also be on a storm basis so that the information can be combined for any given storm frequency series to obtain trap efficiency on a time basis. A trap efficiency value for a period of years can be misleading without also presenting storm data. We should also study a wide range of the important influencing parameters, such as various discharge systems. Similarity of data may obscure the importance of some parameters.

Comment

Reservoir-sediment trap efficiency is a very important area of research because we need to design each reservoir to accomplish specific objectives. Such objectives can be accomplished by knowing how to control the movement of sediment in a reservoir and then carefully designing the reservoir so that it will have the necessary characteristics to control sediment trap efficiency.

In future trap efficiency research, we need to carefully characterize and study: (1) the inflowing water and sediment, (2) the dimensions and configuration of the reservoir storage, and (3) the discharge spillway location and capacity. We should study reservoirs larger than 80 ha drainage area (because of more uniform sedimentation patterns), study and measure reservoir performance on a storm basis, cover a wide range of parameter magnitudes, and focus on soil particles in the silt and clay size ranges. The sand-size particles will probably settle out after 60-90 m of travel in any reservoir. Reservoirs primarily vary in the ability to trap sediment in response to their availability to trap the silt and clay particle sizes.

After being satisfied by or restricted to pure empirical relations for estimating reservoir-sediment trap efficiency, for many years it is enlightning to see the recent research directed more toward the physical processes of sediment entering and moving through a reservoir, with a good balance between field and purely theoretical efforts. Such research should soon enable us to greatly improve our predictions of reservoir-sediment trap efficiency.

215

REFERENCES

American Society of Civil Engineers, 1975, *Sedimentation Engineering*, American Society of Civil Engineers Manuals and Reports on Engineering Practice No. 54.

Beer, C.E., Farnham, C.W., and Heinemann, H.G., 1966, Evaluating sedimentation prediction techniques in Western Iowa, *Transactions, American Society of Agricultural Engineers*, 9, 828-833.

Bondurant, J.A., Brockway, C.E., and Brown, M.J., 1975, Some aspects of sedimentation pond design, in *Proceedings, National Symposium on Urban Hydrology and Sediment Control, University of Kentucky, Lexington*, C-35 - C-41.

Borland, W.M., Reservoir Sedimentation, in *River Mechanics*, ed. Shen, H.W., (Water Resources Publications, Colorado State University, Fort Collins, Colorado), 29-1 - 29-38.

Brown, C.B., 1943, The control of reservoir silting, *US Department of Agriculture, Miscellaneous Publication* No. 21.

Brown, C.B., 1944, Discussion of "Sedimentation in Reservoir by B.J. Witzig, *Transactions, American Society of Civil Engineers*, 109, 1080-1086.

Brune, G.M., 1953, Trap efficiency of reservoirs, *Transactio American Geophysical Union*, 34, 407-418.

Brune, G.M., and Allen, R.E., 1941, A consideration of facto influencing reservoir sedimentation in the Ohio Vall region, *Transactions, American Geophysical Union*, 22 649-655.

Camp, T.R., 1945, Sedimentation and the design of settling tanks, *Proceedings, American Society of Civil Engineers*, 71, 445-486.

Chen, C.N., 1975, Design of sediment retention basins, in *Proceedings, National Symposium on Urban Hydrology and Sediment Control, University of Kentucky, Lexington*, 58-68.

Churchill, M.A., 1948, Discussion of "Analysis and use of reservoir sedimentation data" by L.C. Gottschalk, in *Proceedings, Federal Inter-Agency Sedimentation Conference, Denver, Colorado, 1947*, (US Bureau of Reclamation), 139-140.

Curtis, D.C., and McCuen, R.H., 1977, Design efficiency of stormwater detention basins, *Proceedings, American Society of Civil Engineers, Journal of Water Resourc Planning and Management Division*, 103, 125-140.

216

Dendy, F.E., 1974, Sediment trap efficiency of small reservoirs, *Transactions, American Society of Agricultural Engineers*, 17, 898-908.

Gottschalk, L.C., 1965, Trap efficiency of small flodwater-retarding structures, *ASCE Water Resources Engineer Engineering Conference, Mobile, Alabama*, preprint 147.

Guy, H.P., *et al.*, 1958, Sedimentation yield from small watersheds and its retention by flood retarding structures, *Report of Progress to June 1957*.

Hazen, A., 1904, On sedimentation, *Transactions, American Society of Civil Engineers*, 53, 45-71.

Hazen, A., 1914, Storage to be provided in impounding reservoirs for municipal water supply, *Transactions, American Society of Civil Engineers*, 77, 1539-1640.

Heinemann, H.G., and Reynolds, L.E., 1962, Interim findings on trap efficiency of small flood retarding reservoirs, in *Manual of papers prepared for Joint Sedimentation Workshop conducted by ARS-SCS, Panguitch, Utah*.

Pennell, A.B., and Larson, C.L., 1976, Effects of design factors on sedimentation basin performance, *Paper 76-2020 presented at the ASCE Annual Meeting, Lincoln, Nebraska*.

Rausch, D.L., and Heinemann, H.G., 1975, Controlling reservoir trap efficiency, *Transactions, American Society of Agricultural Engineers*, 18, 1105-1113.

Roehl, J.W., 1975, Procedure - Sediment storage requirements for reservoirs, *US Department of Agriculture, Soil Conservation Service Engineering Division, Geology Technical Release*, No. 12.

Schiebe, F.R., and Dendy, F.E., 1978, Control of water residence time in small reservoirs, *Transactions, American Society of Agricultural Engineers*, 21.

US Department of Agriculture, Agricultural Research Service, 1975, *Present and prospective technology for predicting sediment yields and sources*, US Department of Agriculture, Agricultural Research Service Publication No. ARS-S-40.

Ward, A.D., Haan, C.T., and Barfield, B.J., 1977a, Simulation of the sedimentology of sediment retention basins, *University of Kentucky Water Resources Institute Technical Report 103*.

Ward, A.D., Haan, C.T., and Barfield, B.J., 1977b, The performance of sediment detention structures, in *Proceedings, National Symposium on Urban Hydrology, Hydraulics and Sediment Control, University of Kentucky,* 58-68.

Ward, A.D., Haan, C.T., and Barfield, B.J., 1977c, Prediction of sediment basin performance, *Paper No. 77-2528, presented at the 1977 Winter meeting of ASAE.*